Miserable At Work?
Why?
You Don't Have To Be.™

A practical how-to-guide for intentional fun delivering creative ideation for innovation.

Part of the Breakthrough Thinking™ series from
The Ideation Emporium of Creativity™.

By
George M. Nagle

Foreword by Carlos M. Cardoso

Miserable at work? Why? You don't have to be.™ A practical how-to-guide for intentional fun delivering creative ideation for innovation by George M. Nagle

Published by The Ideation Emporium of Creativity™, LLC
https://www.IdeationEmporium.com

This book, and others, from the Breakthrough Thinking™ series from The Ideation Emporium of Creativity™, are available in printed, electronic, and audio versions.

Miserable at work? Why? You don't' have to be.™ A practical how-to-guide for intentional fun delivering creative ideation for innovation by George M. Nagle;
1st ed.
ISBN: 978-0-9862472-5-5
Library of Congress Control Number: 2022903754
Original 1st print March 2022, Okemos, Michigan, USA

Dedicated to my son Trevor

CONTENTS

FOREWORD III
INTRODUCTION VI

SECTION ONE: LETTING GO 1

CHAPTER ONE -Having Intentional Fun At Work 2
CHAPTER TWO- Being Open And Present 9
CHAPTER THREE- Being Brave Is Simple, But Not Easy 13

SECTION TWO: ENERGY SETUP 19

CHAPTER FOUR- Bring The Energy 20
CHAPTER FIVE- Indicating And Setting Expectations 33
CHAPTER SIX- The Long And The Short 36

SECTION THREE: BREAKING BARRIERS 41

CHAPTER SEVEN- Creativity Barriers, #1 Is You 42
CHAPTER EIGHT- Creativity Barriers, #2 Is Time 45
CHAPTER NINE- Creativity Barriers, #3 Is Indecision By Risk Aversion 49
CHAPTER TEN- Creativity Barriers, #4 Is Language 52
CHAPTER ELEVEN- Creativity Barriers, #5 Is Involving Customers 59
CHAPTER TWELVE- Creativity Barriers, #6 Is Stuck, Diluted & Killed Ideas 80

SECTION FOUR: UNLEARNING FEAR 84

CHAPTER THIRTEEN- Best Ideas And Neuroplasticity 85
CHAPTER FOURTEEN- F.E.A.R. Of F.A.I.L.ing 91
CHAPTER FIFTEEN- Kids -Vs.- Adults 94

SECTION FIVE: CREATIVE IDEATION 97

CHAPTER SIXTEEN- Building Up Ideas 98
CHAPTER SEVENTEEN- The Tools 104
CHAPTER EIGHTEEN- The First Reduction Round 110
CHAPTER NINETEEN- Customer Feedback 113
CHAPTER TWENTY- Time For Transformation 116
ABOUT THE AUTHOR 118

FOREWORD

There are a plethora of business books available on just about every subject imaginable. The business case study books like "Built To Last" by James Collins and Jerry Porras bring out some historic excellence found in various companies. These examples highlight best business practices to form stable organizations. There are also books like "Fast-Track Your Business: A Customer-Centric Approach" by Laura Patterson. Her book includes research aspects from other companies mixed with practical explanations that outline useful tactics for most businesses. Then there is "Miserable At Work? Why? You Don't Have To Be.™," and the Breakthrough Thinking™ series.

If you need to implement change in your organization's innovation, this is a guide to walk people through that transition. It is an enticing way to engage people that almost is a recipe to market-changing creation. I say it's almost a recipe because it is clear that the tools are designed for constant adaptation to your business needs, but more importantly, to your cultural needs.

Earlier in my career, I remember a story from James McNerney, former CEO of 3M and Boeing, as he transitioned from General Electric to 3M. One of the numerous reasons he transitioned to 3M was his vast understanding of the Six Sigma processes from GE. He attempted to take 3M and fit it into the GE box of tools. This frustrated parts of the 3M organization. When he transitioned to Boeing five years later, he had learned to take stock of where Boeing was as a company first. He discovered the culture already had some extraordinary LEAN processes in place in his onboarding assessment. He kept those and added different Six Sigma processes that complemented what was already working.

"Miserable At Work? Why? You Don't Have To Be.™" reminds me of this story as it begins a journey so that you understand where your culture is. Throughout my many business experiences in turning around organizations, the starting point is always assessing the business. You find the 'how' and 'why' of a culture to embrace the positive aspects while removing the less productive ones.

The connection to the people first resonates for any company undergoing changes or needs to transform its' collective thinking. This book does that and helps build bridges for people in their business and personal lives. The tools are presented so that they have an element of improvisation to suit the strengths and needs of the people who are implementing new techniques to grow organically and redefine market spaces.

The focus on people is empowering for transformation. Having spent most of my career with engineers and science-oriented teammates, I found the theme of having fun to achieve breakthroughs in thinking to strike a chord with me. By bringing people back to their original comfort zone of child-like thinking, there is a natural escape from what we have already created for the market through fun and laughter.

"Miserable At Work? Why? You Don't Have To Be.™" is short and to the point. Our author, George Nagle, has provided us with an easy-to-follow guide to achieving new heights in our chosen markets. His extensive training and experience in different industries around innovation delivers proven methods that are universal in their application. The tools and concepts presented describe how to simply and succinctly get out of the rigid adult comfort zones to an open, creative, and adaptable thinking that many of us have forgotten.

I hope you enjoy your voyage into Breakthrough Thinking™ as much as I did. -***Carlos M. Cardoso****, Active Board Member for Black & Decker and Hubbell Inc., Principal of CMPC Advisors, LLC, Former Chairman and CEO of Kennametal, Former Non-Executive Chairman of Garret Motion, and Global Executive.*

INTRODUCTION

Welcome to the Breakthrough Thinking™ series! Thank you for deciding to take this journey with us. The possibilities within you for change are outstandingly astonishing and miraculously inspiring.

We recommend that you read "Miserable At Work? Why? You Don't Have To Be.™" the whole way through the first time before taking action, so you get the entire picture. The picture in your mind will become a map to guide you as you continue with Breakthrough Thinking™. Once you have a map, you can begin to lead yourself and others, regardless of titles, along an amazing adventure.

This book is for anyone feeling stuck, frustrated, exhausted, and perhaps angry about how their life is going. If you want to change you, your relationships, and the markets you serve, while leading the way in innovation, then this book is for you. It is designed to bring the best out of everyone. This could be at work or home, though this book tends to phrase things in the work perceptive.

In our journey together, you will understand the 'why' of how we, as humans, do what we do. With that awareness, you can be open to having intentional fun at work and the benefits that fun bring throughout your life. You will then get a rather detailed map on how to actually accomplish astounding feats within your markets by using simple, rememberable, and fun tools to lead.

Customers are extremely important; so important that we lay out a solid foundation before we start addressing them specifically. We discuss customers a great deal, especially starting in chapter eleven. We invite you to conceptualize everyone as a customer. Please keep this notion in the back of your mind for each section and chapter.

This book has been deliberately kept to a minimum to maximize your time when reading and learning the techniques to unlock the innate creativity that is in all of us. Additionally, it was written to keep the material simple to follow so that anyone can use it.

There are elements of self-discovery throughout the book. Take the time to try on a different point of view with aspects you might disagree with. More importantly, invest time with sections that you feel are incomplete for you. This is a way to widen your perspective from within by engaging your curiosity. Imagination mixed with curiosity delivers creativity.

The part that we can't do for you is using the information inside. Once you do, it will change your life and the lives of those that you share it with. The book is simple, but that doesn't make the discipline of following the steps and ideas any easier.

We hope you enjoy "Miserable At Work? Why? You Don't Have To Be.™," and the other two books, "The Five Letter F Word™" and "How Are Kids Innovating Faster Than You?™," from The Ideation Emporium of Creativity™. Thank you for embarking on this journey with us!

SECTION ONE: LETTING GO

CHAPTER ONE -Having Intentional Fun At Work

"If you do what you love, you will never work a day in your life."

Anonymous

We have all heard that quote or a version of it. Ideally, if you enjoy what you are doing, you won't mind the effort going into the work. This philosophy is often seen as a luxury for a few and not feasible to most. We respectfully disagree and welcome the opportunity to show you how you can realistically achieve happiness in what you are doing and how having fun is incredibly important.

Let's explore the notion that it would be nice to have happiness in our daily working lives, but it just isn't in the cards for us. When you think about that, it feels wrong, but then the social rules you have always operated under say otherwise. However, we need to and should be having intentional and purposeful fun at work. That is why it is the first and most significant aspect of this how-to guide, and where we start in The Ideation Emporium of Creativity™ when we do live sessions with partners.

Have you truly smiled today? Think about that. Has there been just a moment where you authentically smiled and felt that unique vibe? Our first how-to is for you to begin each day with a starting smile within the first hour. Starting your day with something that makes you smile, or laugh, or have fun, is a setup for your entire day that takes just a few minutes. Interact with your children by playing a silly song they will sing or dance to. Maybe go online and learn a new joke or watch some favorite clips. Find your starting smile.

When we are having fun and smiling, time seems to move quickly. For more extroverted individuals, maybe it's having a massive party, while for more introverted individuals, it is enjoying a board game with a few friends after a cozy dinner. When this happens, those moments seem to be almost magical. Usually, you will want to extend your participation in the activities because they feel like they just started. Oddly, we seem to reserve those periods for the weekends or when we don't have work responsibilities the next day.

Addressing the desegregation of when we should be having fun is our second step. In fact, if you are in a work environment where you can't have fun and are not deriving some satisfaction beyond a paycheck, you should find other employment. We spend too much time away from our loved ones to have it be so daunting. Of course, if you happen to be incarcerated, you don't have a choice, but you do in almost every other scenario.

You need to learn how to have fun again, wherever you are. You will naturally draw people to you because fun people act as a magnet towards others. That magnetism is powerful. It will let you start building exceptional teams and maybe even ensembles of people who all want to work with you on whatever you're doing.

Think about the last home improvement project you tackled yourself that required extra hands. Did you call the people you work with to help, or did you call a friend or two? They might be one and the same for some people, which is fantastic. For others, they certainly called their friends. We call them because we trust them, and we want the project to be fun while we do it. By the way, one of the reasons you trust them is because you have fun together.

The most remarkable example of the power of fun in drawing people to you is through children. Children will laugh between 250 to 400 times a day, compared to about 17 times a day for an adult. When was the last time you heard children laughing and secretly wanted to see what was so funny? The pureness of their joy causes you to smile naturally. You may have found what they found funny, but the laughter itself is what is infectious. Many people describe feeling a spark at the sound of children laughing followed by a drop when it has stopped. Their laughter touched our locked away youth, and we miss it, just for a moment, when the laughter ended.

A natural question is, why do we seem to lose that spark of youth that makes us laugh 300 times a day? Part of the answer is maturity mixed with responsibility. The other part is because we start to restrict imagination.

For a moment, picture isolated communities of people in parts of South America or Africa. The adults in those communities may not laugh as often as their children, but visitors describe everyone typically as joyous, energetic, and excited about exploring life.

What is the difference? The most glaring difference is the education system. Please, do not get us wrong; the value of education is immense. However, there is a cost to education well beyond the money spent each year. The process is, in its very nature, restrictive. Sit here, do this, stay in the lines, you can't use that word, 2+2 doesn't equal 22, etc. We recognize the power of the arts in education, but even that process is still limiting. We aren't advocating for a different system because we don't have a ready-made suggestion to replace it. The point is to bring awareness to start thinking about how to bring out, at appropriate times, that imaginative, creative, playful, amazing, unique you.

The goal is to bring back some playfulness into our lives. This is in a general sense but also with a purposeful intention at work. Remember that work is a large part of our lives, so it needs to be playful at times. The ability to play with those around us creates fun. As we will discuss later, this allows us to get into our Aspiring Alpha brain where Breakthrough Thinking™ can flourish.

The power of intentional fun releases the stress that we can build up. It engages our child-like thinking so we can innovate like children do to deliver remarkable solutions. The fun does have to be intentional at work. People already know how to find some escape to relieve stress at work in a form of fun, but it usually has a negative effect. These are the gossip groups and cliques that can form that create silos and force disconnection. Intentional fun and playfulness do the opposite. That may sound like an obvious concept given what we just discussed, but when play becomes so ingrained into our culture, you move away from teams in silos to ensembles.

Yes, just like an ensemble in a play. Ensembles members understand that regardless of their role, their priority is to ensure everyone succeeds. They do this because they know everyone else in the ensemble prioritizes making sure they succeed too.

Think about great sports teams. Some of the best teams in history have one or two outstanding players among many average players. When teams have multiple superstars among them, they often struggle because the superstar performers become frustrated and express that they, as an individual, are not getting enough time, touches, opportunities, etc. The difference is when teams perform as an ensemble, they all win.

But how do we get to that point? How do you get to the best team without worrying about the best players? The answer is simple; you get people playing and laughing together.

For our discussion going forward, when we refer to playfulness or having fun consider it to be intentional when thinking in terms of work. That doesn't mean that spontaneous fun is bad or should be excluded. In fact, that is the best type of fun when it can be harnessed into a common goal. It means you have to think about having fun. Given all of the things you have to think about, that should be a pleasant prospect of something to do.

The fun can build bonds with the right mix of positive and negative interactions. Specifically, you get people doing this in a five positive to one negative ratio. This does not mean that all of the interactions are positive.

Dr. John Gottman and Dr. Robert Levenson have studied married couples for decades and found that the 5-to-1 ratio has a phenomenal correlation in the data to a successful partnership. We spend a lot of time with our coworkers. If we were able to employ that positive to negative ratio, along with clear priorities, the ensembles we would form would be unstoppable.

When something negative happens, like a disagreement that feels personal, an annoying habit, not involving the other in a big decision, wasting money, not inviting someone to lunch, etc., it takes at least three or four positive things to offset it. Plus, a little extra for the relationship to stay strong. We need the good feelings in order to weather the bad ones. The correlation demonstrates how powerful negative interactions that often evoke fear, and the secondary emotions that come with it, can be. Though if things are only positive, it can set up a mighty fall if something goes wrong later that may feel impossible to get over. We need to have both positive and negative interactions in balance.

It may sound foreign that a ratio of interaction could create this type of bonding. Yet, consider two seven-year-olds on the playground. Every day they get to play together. They build off of each other's imaginations and become the best of friends. There is give and take in their play, causing the balance we described. One day they have a massive fight about something. They go home and tell their parents how they will never play with that kid again. They describe how horrible they feel because of how the other person wronged them. The next day, when those children return to their homes, they will more than likely describe playing with their best friend again.

The power of playing with people and the chemical releases it gives our minds and bodies is staggering. It can allow us to lower our guard and start to considering changes. When we are having fun, we are just in a better space.

Think about someone you don't like working with. Or maybe as a leader, you know two team members that do not get along, but you need them to work together. When was the last time they laughed together? If the answer is never, you now have the awareness to bring change. It is hard work to make that happen, but the bonds will form when you get five positive interactions to one negative interaction as a constant.

Will there be great individual performers that will still be the lone wolf no matter what you do? Absolutely. Bring them awareness, help, guidance, and laughter. If they are still a constant, deliberate pain to the organization that is trying to move forward as one, get rid of them. No amount of talent is worth the continuous derailment of the whole.

CHAPTER TWO- Being Open And Present

"When we Open up to the Present, we get the gift of the future."

Two other incredibly significant aspects go along with the power of having fun to bring out your innate creativity. They are being open and present. Let's briefly explore each of these.

Being open can be scary. It can invite change and the perception of more things you HAVE to do, which means giving up something you WANT to do. However, it doesn't have to be, and it shouldn't be. Think about trying on a new coat. You are open to the potential that it is a good fit, feels comfortable, and looks good. If it doesn't do those things for you, you simply take it off because the feeling doesn't justify the price. You set it down and move on no worse for wear.

In our journey throughout this guide, we ask that you do that. Try things. Give them a fair shot without any predetermined notions. Some aspects will work for you; some may not. Embrace what works for you, but remember, if you get stuck later on in the future, come back and try those things that didn't work for you before. They will feel uncomfortable.

Being uncomfortable leads to breakthroughs you would never imagine. When we are out of our comfort zones, we are experiencing things in different ways. It is the differences in those experiences that can unlock us, as you will discover throughout this journey.

Being present is about staying in the moment. It is about committing yourself to what is in front of you. And, more importantly, who is in front of you. Breakthrough Thinking™ requires you in these moments to be with the team, not lost in emails, or texts, or phone calls. It is easy to get distracted and think about our mounting work. However, being present is a lot easier when you start having fun.

If you are a leader in your company or the one organizing the team to drive the session, you can cause some damage by not being fully present. You are indicating that the topic at hand isn't important to you. You may be displaying, unintentionally, that the input from others is not worth your time. The fastest way to turn something that everyone wants to work on into something people will dread is when a leader doesn't actively display the value they feel for it.

When you are present, you are engaged in what is happening in front of you. If you know you get distracted by a message on your phone, then turn it off or leave it where you can't see it or hear it. If a presentation is going on, and it hasn't kept your attention, ask yourself two questions. "*Why am I attending this presentation?*" and "*What value am I bringing to this presentation?*" They will change your perspective about the material and people in front of you. This is what leaders do when they are present.

An important distinction is needed here. Some of you may be a leader regardless of a having title or not, but be aware if you have a title that you might not be a leader. A leader does not mean a manager, supervisor, vice president, or even C.E.O. Those titles allow people to tell others what to do. Conversely, a leader empowers people to take ownership. A leader inspires others by giving a vision of what is needed, then, potentially, taking a step back at times to let others flourish. The quickest way for those with a title to lose respect as a leader is by not being present.

Being present as a leader means you also can drastically sway a conversation with a team. The cliché adage of "*with great power comes great responsibility*" is true. Your responsibility is to recognize the power of your words, especially if your culture is new or undergoing change. Despite your intentions, what you say can have unintended restrictions.

One further consideration about being present involves your customers, which we discuss in a great deal starting in section three. Have you ever been with a customer, and they weren't engaged with what you were presenting? How did that feel? Pause for a moment and recognize that every person you interact with is a customer because they are all human. What you and every other person is selling constantly are their views, hopes, needs, wants, and inspirations. By engaging with those around you, no matter who they are, you will find it much easier to display your value regardless of who is purchasing.

The other part of being present is to have focus. The focus needed here is about being aware of you, those around you, and those you will create solutions for. Read "The Five Letter F

Word™" if you are interested in more on how focus can transform not only a project but an entire organization.

CHAPTER THREE- Being Brave Is Simple, But Not Easy

"The opposite of bravery in our society is not cowardice; it is conformity."

Rollo May/Earl Nightengale/ Robert Anthony

Similar to every species on this planet, our first priority is survival. After primal needs, we are social creatures. We have a need to fit in and be accepted by others like many other mammals do. However, something distinctly human is the need to be correct. When you put those three attributes (survival, fitting in, and being right) into focus, you bring awareness to a few keys to unlocking your organization, team, family, and, most importantly, yourself. We will examine those aspects and demonstrate how mastering your knowledge of these can produce Breakthrough Thinking™.

Let's consider for a moment this need to be correct, what some might term as pride or ego. It is a powerful emotion and based on the most powerful emotion of all, fear. The fear is how people will perceive us in a negative manner. When we are wrong, we may think we appear weak.

When others that you interact with, at least semi-regularly, make an error, do you automatically think less of them? Typically, the answer is no. You know that being incorrect is just that, being wrong. It is a mistake. Or maybe it's simply not knowing an answer. Generally, we don't dwell on the mistakes others make unless they are wrong a great deal of the time. So why do we focus on it internally when we are wrong?

Whether you are a leader or believe yourself to be, being wrong goes against what we think are society's accepted leader traits. It is that perception that we self-impose that creates that need to be right. When that need turns into pride or ego, the very thing you are attempting to avoid becomes reality for others. You are perceived as only being about you and your correctness. That perception can weaken trust that you have the ability to accept other views, which are needed to be part of a human social group that isn't a cult.

There is a flip side to this thought. What happens when we are in a larger group and have a different solution than others? Consider the old classroom example of being 14-years-old in a math lesson. Everyone has the answer the teacher does, except for you. You quickly double-check the math and can't find your error. Would you raise your hand and say that everyone else is wrong? Or maybe raise your hand and say you have a different answer? Or maybe ask for the problem to be demonstrated?

The vast majority of people would do none of those things. They would pretend they got it just like everyone else. They fear being incorrect or maybe being seen as a standout in the group. They are willing to accept group think.

Then you have the people that can't accept they are wrong. They may argue in the moment or wait until after the class to prove their superior ability.

An even smaller group wants to be different so that they are not part of the massive group. They may say something just to stand out. This is still being part of a group, as you will see.

Then there is the smallest group that wants to understand and can accept that they may be in error.

Why is it that most just go with the large group? Over the centuries, we have manifested our survival based on being part of a community. This manifestation is not surprising given it provides numerous benefits to us, such as shared resources, protection in numbers, readily available mating partners, and the like. And this social survival is not unique to humans. It is, however, essential to understand the social and community bias of why we do some things and why it makes change difficult.

It is also necessary to understand that being right usually does not matter when you form a new idea and the team to go with it. What does matter is creating the space where everyone feels they can express their view and not be excluded for doing so. To do that, you must bring awareness and actively work to overcome the individual perception of what that looks like.

You need to empower most people to share a different answer or even an unpopular opinion, as it goes against their nature. For those that need to be right, you have to help them check that ego at the door and demonstrate diversity in thought has nothing to do with right and wrong. For those that want to stand out and be different, you need to encourage them to be themselves but still contribute. Then, for those that want to understand the 'why,' you need to make sure they are empowered and inspired. They often are the ones that can be the glue to help bond others when working in teams because they naturally want to understand why someone thinks or feels the way they do.

When you instill this bravery in everyone, it gives them the ability to challenge ideas through seeking to understand. Ideas, individuals, teams, ensembles, and communities transform when people feel safe in overcoming the natural feeling of butterflies in their stomach caused by stepping out of their comfort zones. Perhaps your community is your company, but what if it were more extensive than that; how amazing could that be?

So, how do we allow those individuals to be brave? You need to demonstrate, actively, they are safe. The fastest way to do that is to have fun with them. When we play together, it creates an energy that demonstrates safety, trust, and comfort with each other. The simplest way to play together is through laughter, and we will talk about that throughout this book.

Perhaps the most challenging group are those that want to be different just to be different. You can see an extreme example of this in Dr. Theodore Kaczynski, better known as the Unabomber. He claims he has no desire to be part of society and is not concerned with society's thoughts. That is still a method of wanting to look good, so you can be accepted, but in a different light.

It is reported he requests to be addressed as a doctor, which is a title he earned in mathematics. That type of insistence is usually about looking good because his brilliance sets him apart from others. However, this is not unique to him.

His uniqueness, and why he is being used as an example, is because there is a group he wants to be highly accepted in. He wants to be a beacon to the hermits and extremists that want to break down the technological and industrial movements of the world. We know that he is serious about this because he refused interviews except for one by a woman that shared some of his views and praised his manifesto. Also, during his trial, he was determined to not take a defense of insanity because he felt it would diminish his manifesto if he did. He was worried that if he looked mentally weak, it would have a negative reflection on his work.

We certainly not advocating that you bring a terrorist into the fold. He is used to show that even the extremist of the world want to belong. If a deliberate standout is already part of your community, it is usually for a good reason. You need to break down their barriers, so they are included.

How do you break those barriers with those that just want to be different? The answer is slowly and one-on-one. There is no shortcut here. It will take time, focus, attention, energy, absolutely no ego, and most importantly, authenticity. In fairness, authenticity will need to shine through for every person. Without it, no one will take anything you do seriously. They will not invest themselves in anything you are working on.

Being genuinely authentic is challenging. It forces us to leave the socially acceptable comfort zone. The comfort zone is a form of homeostasis, which is the body trying to stay at a status quo. Comfort, in many ways, is a drug. We do have chemicals released into our system when we feel comfortable, creating a form addiction. It is one of the challenges to change the status quo.

It will take you being ready to change and actively demonstrating that change isn't really that scary to help others break that addiction. Be brave, be authentic and be the inspiration for others by doing. By doing the things outlined in this book, even if they feel silly, you will encourage others.

Remember that if the only way you can lead is by a title or ego, that makes you a manager, not a leader. In the words of Steve Jobs,

> *"Management is about persuading people to do things they do not want to do, while leadership is about inspiring people to do things they never thought they could."*

Show people bravery by standing up for everyone to win, and you will too.

SECTION TWO: ENERGY SETUP

CHAPTER FOUR- Bring The Energy

"Our energy is in proportion to the resistance that it meets. We attempt nothing great from a sense of the difficulties we have to encounter."

William Hazlitt

This is a how-to guide, so it's time we begin giving you practical things you can actually do. One of the first tools we want to provide you with are ways to have fun, laugh, and bring energy to your teams. This list doesn't even begin to scratch the surface of all of the energizers or ice breakers out there.

It does, however, give you some of the 'why' and 'where' you would use different ones. As you will see, the various energizers that you use can be a way to indicate to your team or group beyond the innate fun what the purpose of the day is. We indicate if these are good for an expansion when you want to create enormous possibilities or reduction when you bring the enormous ideas down to meet the needs without diluting the idea. Let's begin.

01. **EXPERTS** This activity takes about ten minutes. It is ideal for groups of six or more. It is general in its application and meant to get laughing and Breakthrough Thinking™ to take place. Good for expansion and reduction. Here is how to do it:

A. Pick three outlandish things you feel would never be but aren't entirely out of the world of realism. Some examples: a honeybee sex therapist, fashion designer of invisible clothing, designer of parachutes for elephants, retirement worker for bacteria, financial advisor for rich pets, matchmaker for pirates, language expert for trees, etc. Have fun making up some of your own. This can be an energizer too.
B. Break your participants as best you can into groups of three.
C. Have one participant be the expert, and the other two are reporters. They are going to interview the expert. Make sure the expert is confident that no matter what they say, they are right!
D. After a minute or two, when you get some good laughter going, you know it is time to switch. Have the participants switch roles. The new expert is an expert on a different topic than the prior expert.
E. Repeat until each person has been an expert.

02. **SILENT ORGANIZATION** This activity takes about ten to fifteen minutes total. It is ideal for groups of ten or more. This is useful for bringing awareness to how we communicate and language barriers. Suitable for expansion and reduction. Here is how to do it:

A. Ask participants to stand and leave everything at their seats.
B. Break your participants into equal-sized groups. A group should not be more than ten people.
C. Tell them they are not allowed to use verbal communication. They are not allowed to retrieve anything from their seats or use writing utensils or electronic devices. They are free to use their hands to communicate.
D. Tell each group they must get organized in one single line going from oldest to youngest. They need to get it down to the day.
E. After a few minutes, push them along to finish. This mimics how poor communication can get worse when we are rushed.
F. They compete against the other groups to see who can do it first. Tell them to sound off their age when they think they have it. If there are older people participating, this activity can actually help open them up to the group.

03. **BIGGER, BIGGER, and BIGGER** This activity takes about ten to twelve minutes. It is ideal for groups of six to ten participants. Use this energizer when you want the group to start thinking big and aspirational. Good for expansion. Here is how to do it:

A. Have the participants stand in a circle.
B. Person 1 will say to the person to their left, "*Here is a gift just for you.*"
C. Person 2 will say, "*Thank you for X gift.*"
D. Person 1 has to explain why they got person 2 that gift.
E. Once that is done, repeat with the next person EXCEPT that the next gift must be physically bigger than the prior.
F. The game ends when the last person has gone OR when you get to the universe.

04. **RIPPING IT UP** This activity takes about two minutes plus one minute for each participant. This is good for introducing new people into groups and demonstrating that it is OK to F.A.I.L. Good for expansion and reduction. Here's how to do it:

A. Take a scrap sheet or half sheet of paper.
B. Have participants hold the paper behind their back in both hands.
C. They have sixty seconds to, without looking, rip into the paper their favorite animated character.

D. Their characters won't be correct, but that is the point! Have them show and tell everyone what they were trying to do and why it is their favorite character and introduce themselves.

05. **TAKE WHAT YOU NEED** This activity takes about ten minutes. It is ideal for groups of six or more. It is general in its application and meant to get laughing and Breakthrough Thinking™ to take place. Good for reduction. Here is how to do it:

A. Get a fresh roll of toilet paper.
B. Ask each participant, "*Please take as much as you need*." Do not provide any other instructions beyond that.
C. After each participant has taken however much they need, tell them they need to share one fact about themselves for each square they took. The more unknown the fact, the better. If they took none, they need to share one fact no one else knows about them in the room.

06. **GET TO 21** This activity takes five to ten minutes. It is ideal for groups of four to twelve. The focus for this activity is to promote taking action when others aren't sure if they should, learning from each other, and listening. Suitable for expansion and reduction. Here is how to do it:

A. The group shuts their eyes.

B. Then, one person at a time needs to count out a number starting at one. A person that just said a number can't give the next in the sequence. Anyone in the group can speak.
C. If two people say the exact number simultaneously, you start over again at one.
D. If a number is skipped, you have to start over again as a group at one.
E. Each person has to give at least one number. If the group gets to 21 and someone didn't, they start over but now have to get to 25.

07. **MIRROR** This activity takes about two minutes. This is a pair exercise. It is used to get people comfortable with leading and following. Good for expansion and reduction. Here is how to do it:

A. Have participants break into pairs.
B. Have the participants face each other and decide who is leading first.
C. The first leader can make any face or hand gestures they want, nothing rude, and the partner must mimic it as fast as they can for thirty seconds.
D. After thirty seconds, they switch roles.

08. **CLOSE & FAR** This takes about ten to fourteen minutes. This is for a large group and needs a large area. This is about

observing other people and what body language can tell us. Good for reduction. Here is how to do it:

A. Have everyone select a person they will try to be close to and one they will try to stay away from.
B. Have everyone start walking around in a big mass, not running, for three minutes.
C. See if people can figure out who was trying to stay close to them and who was trying to stay away from them.

09. **YOU THEN ME** This activity takes about four minutes. This is done in pairs. This exercise is great for collaboration and learning how to build together on each other's ideas outside of language. Good for expansion. Here's how to do it:

A. Pair up two participants and tell them they can't communicate in any way.
B. Have them draw a creature together, one line at a time. Person 1 does a line, then Person 2.
C. They go back and forth, one line at a time, for sixty seconds.
D. The participant on the left has to name it without telling the partner. The partner on the right has to describe it when they present it after hearing the name for the first time during the presentation.

10. **ALPHABET STORYLINE** This activity takes about fifteen minutes, depending on the number of participants. It is ideal for groups of five to thirteen. This emphasizes active listening and building on each other's ideas. Good for reduction. Here's how to do it:

A. Assign two different letters to each person. Writing them on a post-it note can make this easier, so there is no duplication.
B. Each person must come up with one word, starting with the letters they were assigned. One of the words has to be a verb and the other word a noun. Example: If they got "A" and "V," they could have a verb as Ascend and a noun as Victoria.
C. Get someone to create a single sentence using at least one of their words they wrote down. It needs to be simple as they will repeat it a lot.
D. After the first person has said their sentence, the following volunteer must create a sentence using one of their two words, but it has to go with the first person's sentence. Any new sentence can be at the start or at the end of the current story, but not in the middle.
E. Repeat with each person going at the beginning or at the end of the current story using one of their words. The story can take odd directions, but it must make sense; that is why everyone is repeating their lines continuously for each new person added.

11. **STORYLINE** This activity takes about ten minutes. It is ideal for groups of six or more. It is general in its application and meant to get laughing and Breakthrough Thinking™ to take place. Good for expansion and reduction. Here's how to do it:

A. Similar to the Alphabet Storyline, but people can just make up any sentence they want as long as it fits with the others that went before them to form a story. They can also jump in anywhere in the story they want.
B. Have them physically stand and insert themselves in the line.

12. **SHAPES** This activity takes about three minutes. It is ideal for groups of six or more. It is general in its application and meant to get laughing and Breakthrough Thinking™ to take place. Suitable for expansion and reduction. Here is how to do it:

A. Point your index fingers straight out. Simultaneously with both fingers, draw a triangle, then square, then a circle in the air.
B. Now one hand does triangle and the other does square, then switch one for circle.

13. **LEAN IN & OUT** This activity takes about five minutes. It is ideal for any number of people. If you are in person, you

can jump instead of leaning. This is like Simon Says, but designed to get endorphins up that can help create bonding. It is general in its application and meant to encourage laughing, but it also gets people moving. Good for expansion and reduction. Here is how to do it:

A. Round one: Do what I say as I say it- Lean In, lean out, lean right, lean left, etc.
B. Round two: Do what I say AND say what I say.
C. Round three: Do the opposite of what I say BUT say what I say.

14. **TOUCH BLUE** This activity takes about ten minutes. It is ideal for groups of six or more. It gets everyone to contribute and explore their creativity. Good for reduction. Here's how to do it:

A. Each round lasts for ten seconds. They get one point for each successful round they complete.
B. Begin by having the facilitator call out something that each participant has to touch. If this is a web meeting, they also need to show it on screen. An example might be "*touch blue,*" or "*touch something warm,*" or "*touch something soft.*"
C. This might be something on their desk, an item of clothing, or something they have to go and find on their bookshelf. The last person to find an object in the allotted amount of time then must select the following thing to touch.

D. Continue to play until each participant had the chance to select a thing to touch. If you want to up the ante, introduce a second attribute to the mix. "*Touch something blue and old,*" or "*touch something red that you can eat.*"

E. Debrief by identifying some of the group's favorite responses and items by encouraging the owner to talk a little about why they selected the item. The points probably will be forgotten, but if someone asks, that can be an indication of someone that may need help of letting go of being correct.

15. **STORY DICE** This activity takes four minutes plus one minute per person. It is ideal for groups of four to eight. This is designed to garner different thoughts about the same thing and bring out diversity and confidence in expressing our opinions. Good for expansion and reduction. Here is how to do it:

A. Go to Storydice.com.
B. You will see that you can electronically roll several dice and random pictures appear.
C. Each person needs to create a story using the dice in order of the symbols.
D. They have three minutes to make the best story they can.
E. Each person reads out the story they made.

16. **FLAMINGO** This activity takes about five minutes. It is ideal for any number of people, even just an individual. This is designed to remind us about balance and to help us focus on the big things that matter the most. Good for reduction. Here's how to do it:

A. Have everyone stand up with nothing in their hands and a little space around them.
B. Explain that this is a balance exercise meant to remind everyone that we all have imbalance sometimes, which is OK.
C. Explain each time they fall to yell "*flamingo,*" allow themselves a quick smile or laugh and try again. After an individual yells "*flamingo*" three times, they must do the chicken flap with their arms three times and have a seat.
D. When ready, have them stand on one foot. After about fifteen seconds, make it a little bit more challenging.
    i. Raise the knee of the leg they aren't standing on.
    ii. Raise the foot toward the rear.
    iii. Extend the leg out.
    iv. Extend the arms out wide and the leg up higher.
    v. Do Warrior III, standing on one leg, arms straight ahead and flat like they are flying.

E. For those who make it to this point, ask them to start thinking about the top three priorities in the project. This will indicate where their thinking is on the project as they try to maintain physical balance while engaging their brain.

If a meeting is going to be over ninety minutes, take a five-minute break. When everyone comes back, do another energizer. Keep the people laughing and energized. The returns will be stratospheric.

CHAPTER FIVE- Indicating And Setting Expectations

"People can't live up to an expectation that they are unaware of having been set for them."

Rory Vaden

Despite the criticalness of indicating what is expected, it is often overlooked. If you skip the indicating of your intentions, it can disrupt what you are trying to do. Along those same lines, if you set the wrong expectations, you can cause results that derail your objectives and set the team up to not succeed.

The old adage of *"be careful what you ask for because you just might get it"* is part of indicating. If you ask someone in an email, for example, "*What do you think about this PowerPoint I need to present to upper management*?" What are they going to tell you? Chances are they will judge whatever it is you sent them and tell you how they would do it. Perhaps that is what you wanted, but most often, what we wanted was someone to help us make it better.

A better way to indicate what you were looking for would be to actually say, "*Please read this over. I need to convince upper management we need to do this. Can you help me make this better*?" This indicates a few things; your final objective, that you are asking for help, and you value their opinion.

The prior example may feel simple, and it is. However, it is powerful too. When we take just a few minutes to think about what types of signals we are putting out there for others, we gain a new perspective. That new perspective comes from their point of view, not ours. When you put others first, they will want to work with you to solve a problem because you have demonstrated you value them. You are actively being present with them, for them. They will naturally want to do the same for you.

Here is another idea that may sound a little radical. Stop inviting people to presentations. Start asking them to see where you are on the journey to the solution you are trying to reach. Avoid PowerPoint as much as you can, especially with decision-makers, such as upper management or even customers. Post things around the room as you go and walk as much as possible. Make it a physical journey they see you taking, and they are being brought along. You will still have to do the PowerPoint presentation for later reference, but it becomes a supplement to what you are trying to achieve.

You may feel overly uncomfortable the first time you do it, but you certainly will have the room's attention. This is your opportunity to bring real engagement to the team because now you are having a conversation.

Given the number of meetings that take place online now, this is a big challenge to pull off. A little imagination can go a long way here. If a slide has nothing but words, there is no need to show it. Show yourself on the screen talking with them. If there are numbers, only show the most important ones and use size forty-eight font. If you have pictures or videos, absolutely show those.

When you can do live meetings, make every attempt to avoid having a physical barrier between you and the person you need to approve or sign up for what you're trying to achieve. The barrier, like a table, indicates that you are separate when you need to be next to them building together. When you are next to them, it subtly suggests that you are on the same side.

A final consideration is to be aware of who you are inviting to meetings. Limiting the number of individuals in a meeting allows for more voices to be heard that are impactful. This delivers more value for everyone. Also, think about the personality types of those you invite. When people walk into a room and see others that they know, they can get an indication of what is happening that day. For example, if you are past the stage of expanding ideas and need to focus on reducing an idea to a target market, do not invite a bunch of big idea people. They will disengage or, worse, side-track the entire meeting.

We used meetings and setting up meetings as a way to demonstrate how to indicate. Indicating is beyond meetings. It is the consideration that you are putting forward to help others get from where they are to where you are. By putting in that consideration, what you are doing is meeting them where they are so that the group can take the voyage together.

CHAPTER SIX- The Long And The Short

"Here, we must run as fast as we can, just to stay in place. And if you wish to go anywhere, you must run twice as fast as that."

The Red Queen in Alice Through the Looking Glass by Lewis Carrol

---

If we were to ask you to just breathe in, you could do that for maybe ten to fifteen seconds on average before you couldn't do it anymore. On the flip side, if all you were permitted to do was breathe out, you wouldn't last very long. The end result to either is you pass out if you try to persist.

That same idea applies to having short-term goals and long-term goals. You must do both to survive. You can't do them simultaneously. They have to happen in tandem with equal importance.

In the world of business, we often fall to one side or the other. Most established companies are focused on the quarter-to-quarter or even month-to-month results. Whereas a new company that has a game-changing product that has a long vision could run out of funding as they try for perfection. Neither company survives for long if a market suddenly changes.

We can't expect individuals to do both short-term and long-term at the same time. It comes down to trying to do more and more with the same resources. We often think that people's maximum thresholds can just adjust if we just push a little more. What really happens is the people involved are getting burned out. The 'do more with the same' mentality creates a detriment, and a vicious death spiral begins.

Perhaps you aren't even aware it is happening. This unconscious slow creeping burnout is the boiled frog syndrome. To boil a frog, you put it in cool water while slowly increasing the temperature. They are cold-blooded and allow the heat to pass through their body if it is not too aggressive. Before they know it, they are cooked.

Some companies are growing with mergers or geographic expansion, which is fine as long as organic growth is still happening. When you see organic growth going down, the first question should be, "*When was the last time we did something to change the industry?*" If the answer is more than, say, three years, you might be on a downtrend. At the same time, your competitors, which we will refer to them as challengers going forward, may have tapped into something new.

At this point, it is vital to make a distinction between three keywords: Creativity, Innovation, and Iteration. Each has a process, and each has a place and a time. When you are trying to work on more significant or perhaps longer-term and meaningful projects, you more than likely want to use the first two. Iterations will be perfect for shorter projects that just give a bump here and there.

Creativity is an idea or what we bring to an idea or solution that makes it more appealing and different. Here's where you want to be expanding ideas and having outlandish aspirations. Bringing final users in on the creativity, and remembering they need to be involved throughout the rest of the process, is critical to success.

Innovation is the process of taking those creative, appealing, and different ideas to create something useable by others. The key here is to make sure the original idea goes through the innovation process and doesn't get diluted. We will talk about this in later sections. It will look like the beginnings of a modified Stage and Gate (a process involving different checkpoints where materials are presented to assure upper management approves of progress) because it is.

Iteration is also a process where we add to an innovation or even decide it is time to retire a product. Think about adding a new ice cream flavor or perhaps new selectable colors in the new year's vehicle model. Even expanding into a new geography could be an iteration. Traditional Stage and Gate tend to be great for these types of changes in the short term.

For a moment, imagine holding a twelve-ounce can in your hand with your arm fully extended. Holding it there for thirty seconds or even sixty seconds is not a challenge for most of us. But what happens when you need to keep it there for three minutes? Or ten minutes? Or even an hour? Regardless, the weight of that can creates stress to the arm. The ongoing stress makes the can feel heavy and brings your arm down. At some point, you will need to rest your arm. The mind is a muscle and needs time to relax, decompress, and become reinvigorated, just like the arm muscles.

The concept of needing to recharge ourselves is often talked about; however, we don't seem to be able to execute on it. Perhaps the inability begins in our professional life, but it can quickly spill over to our home lives too.

Part of the reason we don't recharge is because we do projects that are like holding that can for fifteen minutes or even an hour. We tend to have many of those types of projects going at the same time. Our mental status gets locked into a channel that is all about doing as little as possible to try to get some type of rest. If we are doing the bare minimum, we aren't doing our best. Certainly, we aren't going to feel inspired by what we accomplish.

The U.S. military starts each day with making the bed. The idea behind this is as they start each day, they have already accomplished something. They are aware of the power of accomplishment and embrace it. Is something that small, quick, and seemingly mundane really helping inspire people? The answer is yes for many individuals who recognize it as an accomplishment.

In your projects, even the very long-term ones, you need milestones that trigger business actions and people actions. People need recognition for what they have accomplished. When people do many important and valuable things and don't get feedback, they lose momentum. They feel the mental stress of holding the proverbial can without any positive relief that comes through accomplishment.

It is vital to keep momentum moving forward for people so that they keep projects moving forward. Potentially, a milestone for an individual is a point where they stop working on a project's activities for a while to regain some new insight. That doesn't mean they are no longer part of the ensemble of people working on it; it's just their time to pause. In this way, they can do something short-term before coming back to the long-term.

The number one reason corporations do not successfully do this is because they are trying to do too much. They are not focusing on the big important things. They spread people and resources too thin, as discussed in "The Five Letter F Word™." These actions make Breakthrough Thinking™ nearly impossible because people are running just as fast as they can, just to stay in place.

SECTION THREE: BREAKING BARRIERS

CHAPTER SEVEN- Creativity Barriers, #1 Is You

"To be able to break down the barrier of space between self and other, yet at the same time to be able to maintain it, this seems to be the paradox of creativity."

Marion Milner

As we continue along our journey, it is essential to remember to embrace child-like thinking. Doing so does not mean we are acting childish. It means we are open to exploring and not setting limits too early. However, it is vital to recognize some natural barriers to creativity. Then, step away from them.

For anyone that may have read "The Five Letter F Word™," the Power of 3 and 5 is outstanding. When we break things down into three to five areas or categories, we easily remember them. The Power of 3 and 5 also tends to focus our attention if we confine ourselves to a smaller selection to get things done. In talking about the creative barriers, we will break that rule and talk about six elements. There is one reason that is more extensive than the other five combined.

The first and by far most significant barrier to creativity is **YOU**. That may strike you as odd since we have been going along a path designed to show you are creative. You are very creative, but you are also the single greatest barrier to your own creativity.

Please try out the following thought sequence by pausing after each sentence, and give it consideration on its own merit, and then in relationship to the prior sentences.

Perspective defines all of reality. Perspective is always unique to the individual undergoing the experience of reality. Prior experiences, due to neuroplasticity, create a bias towards anything new and will redefine it to the individual. That bias is so prevalent that life is 10% of what factually happens and 90% of how we interpret what occurred explicitly to us. Ultimately this translates into reality not existing as a one-size-fits-all.

The argument to that is we wouldn't function as a society if reality weren't something we all recognize. Keeping in mind the social aspect of our nature, it is more than likely that wanting to be accepted, and manifesting the same perspective as those around us, is how we collectively define reality. The larger the group, the more we do not like to be the outlier.

For awareness, seven factors influence which groups we choose to interact with: sex, family, peers, religion, physical environment, media, and money. Each factor influences us in some form or fashion. The degree to which these factors influence us is unique to the individual. These factors are explored more in "The Five Letter F Word™" but are important here.

These factors help form our natural bias. They are the internal monologue that is constantly on. The consistent input from yourself creates a barrier to the outside world and has allowed you to make it this far in life. That same barrier is holding you back from Breakthrough Thinking ™ because it creates assumptions to define a reality that is not there.

Your bias may present itself in a lot of ways. Here are several thoughts or sayings we all have said at some point that are indications of bias thinking.

We have a process for this for a reason.
We have always done it this way.
We have already tried that.
Doesn't that person ever stop talking?
Can that person try to have an original thought?
That will cost too much.
That is another person's responsibility.
I didn't ask for your views or opinions.
Upper management won't listen, so why try?
Nothing that person says ever makes any sense.
Does anyone else do anything around here?
What do they actually do during the day?
That is not in the budget.
We don't have time for that.

There are, of course, many more. The path to Breakthrough Thinking™ isn't about stopping our bias thoughts or internal monologue. That is nearly impossible. It is about recognizing it and doing something about it. It will require bravery and practice, but that is how we learn and grow.

CHAPTER EIGHT- Creativity Barriers, #2 Is Time

"Time is our greatest and most scarce resource. Ultimately, the only thing that matters is how we spend what time we have because we can't get anymore."

Creativity needs you to take time to think. That may not, at first, feel right, but by the end of this book, you will absolutely understand the profoundness of that need.

The next thought most people have when they hear they need to take time and think is something like, "*I don't have time to eat lunch or get to the 150 unread messages in my inbox, let alone to take time to think*." We have all been there. But if you do this multiple times a week, or you are constantly living for the weekend, or so tired at the end of a shift you have no energy to spend with others outside of work, your issue is not time. It is how you spend it.

The work that you do is not unique to you. As an employee or even business owner, you are replaceable. An ad to replace you at work can be almost instantly generated. There is no replacement for you at home, and you cannot be replaced as a family member or friend. You only get so much time; make sure you are spending it the best way that you can. To be clear, this means spending time outside of work.

When you prioritize living your life, your work life will get better. Though make no mistake, you absolutely will need the support of your team and work culture for this to happen. In some workspaces, it is scary to have these types of conversations. Still, if you AND your company are serious about Breakthrough Thinking™, this conversation will be more natural. If your work is not open to this or you suspect they are not, place energy into finding a new position. Then have the conversation to see if they will put things immediately into place to keep you. This shouldn't take three to six months. In fact, if they don't immediately do it, they won't do it at all. We flush this issue out in "The Five Letter F Word™."

The question you may have is, how do you find time? The answer is you don't because you already have it. The only thing you have that is truly yours is time. The problem is you are too diluted, meaning you simply need to stop doing some things. That is a radical notion, but it is the only logical one. Take a lunch, do not take work home after being at the office all day, and absolutely sign out when you are on holidays, vacations, or just days off. Suppose your company is so short-staffed that they can't resolve issues without you. In that case, that speaks volumes about their abilities to innovate, and this book is not for them.

Taking time to think doesn't require a lot of time. When you become practiced at it, you will find five to fifteen minutes a day is plenty. It is your time to just think. Don't be doing something else like exercising. If you are exercising, focus on that. Your focus should be on what you are doing. Will you get ideas while doing other things, especially when you start to destress? Absolutely. However, that is an unintentional thinking but perfectly acceptable.

Find time to think about something you find exciting or read something unrelated to your job. This intentional diversity of thought will be instrumental in unleashing that tremendous creativity you have inside of you. Set a reminder for yourself on your phone so that you actually do this. If you have a smartwatch, it might already have the functionality built into it.

This isn't all just theoretical. We know this works because at The Ideation Emporium of Creativity™, we have done all of these things. It was scary at first and felt like a waste of time. It absolutely wasn't and continues to be the best investment we can make.

As a leader, there are five specific areas you can focus on to help people use their time in ways that will make them feel better, more inspired, and, frankly, happier. These topics have consistently come up as things people wish they had more time for that have a connection to work.

01. **BASIC SELFCARE**- having time to eat lunch without working, the time to use the restroom without leaving a meeting, or time to do 30 minutes of exercising a day.

02. **DEVELOPMENT & GROWTH**- having time to learn a new subject, a new area, a new skill set, or a new language.

03. **ESTABLISH RELATIONSHIPS WITH LEADERS & COWORKERS**- having the time to form more of a connection and understand why they do what they do.

04. **HELP OTHERS IN THEIR COMMUNITY**- to have time to give back to their communities that support the areas around their homes and be seen as a person working for a company invested in the local community like they are.

05. **BE PART OF A SIGNIFICANT PROJECT**- to have time to contribute to something meaningful for the team to push the company forward that they can take pride in showing to their family and friends.

CHAPTER NINE- Creativity Barriers, #3 Is Indecision By Risk Aversion

---

"Stagnation is death. If you don't change or take some risk, you die. It's that simple. It's that scary."

Leonard Sweet

In 2002, former Secretary of State, Chairman of the Joint Chiefs of Staff, and four-star General Colin Powell spoke to a small group of twenty people in Pittsburgh, PA. He imparted many words of wisdom that day about leadership. One of the most striking things he said that day to us was how the military makes decisions:

> *"It is the 40 to 70% rule. You need to have at least 40% of the information, but no more than 70%, in order to make a tough decision. Depending on the urgency and severity of the situation, such as people's lives at risk from snipers, you go as soon as you are at 40%. The risk of fast, unbiased information coming in that all are indicating a course of action is correct at that point is rarely proven wrong with more information. If you find that you have a tough decision*

that isn't as time-sensitive as I just mentioned, and you have gotten 70% of the potential information indicating a course of action, you go. First, the remaining 30%, even if it contradicts the 70%, is most likely not changing the course of action. There is no such thing as a perfect choice when making a tough decision. But waiting for that 30% is a time sink, and the stagnation it causes can be deadly."

Fear is by far the most powerful emotion. It holds us captive in the moment. When it comes to making a tough decision, it can fool us into thinking the choices we are about to make can't be undone. That simply is not true. The truth is every time we make a decision, we must continue to make that decision as we move along a process. Rarely are we so locked into a tough business decision that we have no recourse to correct it if we made the wrong choice. It isn't like you are making life and death decisions while ideating and innovating.

The sign of a healthy organization is one that actually takes some risk. It is calculated risk for sure. Suppose your organization is successful in everything they do within a market with more than 50 customers. If that is the case, they are not going to be around very long if there is even a slight shift in that market. They are only iterating and not leading the market. They are a commodity player.

New things are risky, and breakthroughs are even scarier. You may not be successful on nine out of ten things you try, but that one you are successful on will easily cover the other nine.

If your innovation process has reasonable robustness, the nine unsuccessful things won't actually make it into the market.

You will have the market involved early and often enough that they will kill it for you. So the question is, out of things that do make it through launch, with robust customer involvement during the innovation process, what is an acceptable amount of unsuccess? That is up to your organization, but the 80-20 rule is a good measure, like in many business aspects. If you have 80% of your solutions being a breakthrough success, you are dominating the market.

The other 20% will be lessons learned by trying something new. The reason it didn't succeed could be due to timing, acceptance, unforeseen technology shortages or shifts, regulation changes, or your challengers in the market-beating you to the finish line.

Newer organizations may argue that they can't afford two things, let alone nine, to be unsuccessful in the marketplace. That is true, and it is also why many businesses go bankrupt in the first year. These organizations try to make it perfect and right immediately. That rarely works. You need to prototype, talk with people, and try things out before going to the market. When you are 40% to 70% there, you go to the market.

When newer organizations come into a market, they tend to be wildly changing things. Early adopters don't expect perfection. They purchase because they are inspired by the new solution. You need to start bringing in some revenue, and it is time to get some short-term focus in. The revenue is more likely going back into the next version of the product, and that is fine. You improve as you go, but you still go.

CHAPTER TEN- Creativity Barriers, #4 Is Language

"Americans and British are one people separated by a common language."

Winston Churchill

Language is a potent invention. It has the power to liberate as well as confine both the speaker and the listener. It is not bound by time, dialect, symbols, or even words; however, the context does restrict the interpretation even in the same language. This restriction is then amplified by our own bias and understanding or definitions of words. A whole new layer of complexity is added when translations take place. In a simpler view, language is almost a living thing that is ever-changing.

Here is a very quick example of what we mean. Did you know that in the English dictionary that nothing beings with the letter 'n' and ends with the letter 'g'? You may be chuckling at this point, or you may need to go back and re-read that question. The answer is yes because the word NothinG begins with the letter 'n and ends with the letter 'g.' The interpretation of the context is both the empowering factor by having the answer and the restrictor because of the word's definition.

You have an opportunity to be honest with yourself about your need to be right. Did you get the answer right? It doesn't matter if you did or not. What matters is your truthfulness to yourself about if you solved it or not. The vast majority of people don't get it right the first time. We didn't either. If your reaction was anything but a smile or a little chuckle, you may need some extra focus on letting go of being correct.

Dr. Lera Boroditsky, a cognitive scientist, delivers a phenomenal T.E.D. Talk about how language influences thoughts. In the United States and many native English-speaking countries, it is common to describe directions with right and left words. In other cultures and languages, they do not use those words in the direction scope. They describe everything in terms of compass points. You may ask why they do that. For them, the physical environment is in deep forest, and by implementing this constant reference to north, south, east, and west, they are never lost. It is also how they describe time as being locked to the environment.

Even the way we describe objects is confined by the influence of language on culture. A great example is how Germans and the Spanish describe a bridge. Germans use a feminine article assignment, and in Spanish, the article assignment is masculine. If a Spanish speaker describes a bridge, words such as sturdy, strong, or spanning are more likely to be used. In German, words like sophisticated, beautiful, and supportive are more commonplace.

As discussed, the bias we all have also influences what we hear the other person say, which may not be what they actually said or even meant. A conversation could go like this when we discover the more obvious bias at play.

Person 1 asks, *"Didn't I ask you to have that to me by noon today when I called you yesterday?"*
Person 2 replies, *"Yeah, but I figured you just meant by the end of the day would be OK because that is usually when you ask for things. You shoulda told me why it had to be by noon; then I would had it done."*

This example also illustrates a different part of language in branding, specifically personal branding. There are three primary ways to view branding; corporate facing the outside world, internal understanding of the culture brand, and our personal brand and what people think about you. We are going to touch on each of these briefly.

Corporate branding is that intangible understanding of a company that comes to mind. There are familiar icons such as Coke, Disney, or Nike. We tend to covet established brands that are usually very selective about the language they use. This is especially true in today's social media, that uses a very powerful magnifying glass of scrutiny that shows no signs of lessening. Good brands are culturally rich and always, for every region they are in, crafted by a native speaker. Corporations that do that poorly can fall flat on their faces.

Two quick examples. First is when companies name a product or solution with an acronym. Those often do not translate well and shouldn't be used. Secondly, a more practical example would be the Ford Pinto in the '70s. When they tried to take that vehicle to Brazil, no one would purchase it. One of the definitions of pinto is 'small penis' in Portuguese. At the time, men primarily purchased vehicles, which meant it was *nada de bom* (no good).

The internal brand or culture of the company definitely has a language. Walk into any work environment with more than five employees, and in the first three days, count the number of abbreviations you have never seen. Then, you have the words that mean something completely different inside than they do outside. Those are part of the culture and become how a group manifests itself. The way to recognize the power is if international employees also use the same terms and abbreviations despite the natural language barrier. It becomes unique to the company. That can be a tool to bring people together, but it also can be used the other way.

Our personal brand, as demonstrated earlier, can be everything from the words we use, how people have learned how you use those words, the tone of how they are expressed, and the body language that comes with them. Personal brands are tough to change once truly established. This goes beyond the hill of a first impression, which is typically formed in under two seconds. First impressions have staying power, but when the collective culture shares your brand with everyone, you are against a mountain, not a hill.

That shared communication in a company or culture needs to be thought through when we approach Breakthrough Thinking™. It is a starting point for how we indicate. Communication is also the most significant issue for any company with more than five employees and in business for more than one year. Communication is the bedrock, as discussed; however, who's problem is it to fix when it is divisive?

How do you fix it, you may ask? As we talk about in "The Five Letter F Word™," you start at the top and keep it concise. You say things in three different ways and with consistency. You say it often and to everybody in emails, meetings, conferences, and most importantly, in-person and one-on-one. The higher up you are in a company, the fewer talking points you should have when speaking to the team.

If you aren't sure where to start, a good place would be establishing a common phrase that sets an expectation. Since we are talking about creativity and how it is essential to innovation, and by extension, what we produce, a good place is defining Productive Creativity. This is a modified definition from Disney that many corporations use today.

> ***PRODUCTIVE CREATIVITY****- the Habit of Continually doing things in New ways to make a Positive Difference in our Lives.*

We have deliberately capitalized the starting letters of keywords. At first glance, that definition might feel like it is a contradiction. However, it is perfectly balanced with the concept that life is constantly changing. When it stops changing for an individual, it is because they are dead. The same can be said for a company.

When we focus on making positive differences, we also further clarify it's not just changing for the sake of change. There must be a positive reason behind it; otherwise, individuals will not get behind it. They must see the value of the change before doing anything new. After putting this new definition into practice a few times, it will naturally become a habit for the culture and be woven into the language.

The modification mentioned is that the word 'working' appears before the word 'lives' in the Disney version. That has been removed because work should not be your life. You should be expressing your creativity and be making positive differences as a way of life, not just at work. Breakthrough Thinking™ is a mindset that carries into everything.

We find that the following words/phrases encourage the power of language to bring change. They are listed in order of importance and how often they should be used.

01. I'm sorry
02. I was wrong
03. You are right
04. I don't know
05. Thank you
06. I need your help
07. Please
08. I want to hear your honest thoughts
09. Can I help you?
10. Tell me about a success you had recently

When possible, they should be used in tandem. An example would be apologizing for being late and thanking them for their patience.

The first four demonstrate humility, acknowledging self-awareness, and a lack of ego. If you are a leader, you must show that you are open to your own imperfection, and maybe you lack expertise. A leader doesn't need to be the best at any specific thing on a team. They are there to be the best at bringing out the best in others.

The power of gratitude is life-changing for those that give it. 'Give it' is not a mistake. You come from a different mindset when you seek ways to thank people. You find the positive in life while becoming the positive in others' lives. It makes you stop looking for what you don't have and realize what you have here and now.

The sixth and seventh words in the list display vulnerability. That does not mean you are weak; it is the exact opposite. It takes incredible strength to ask for help and to implore for it. Using these words will naturally break down walls that can come up between organizations, departments, teams, and individuals.

The final three phrases are about the genuine wish to make others know you are interested in them and willing to invest time in them. Displaying genuine care for others means you are getting down to their level without expectations.

These ten words/phrases are magical when you say them authentically in transforming yourself and those around you. "The Go-Giver," by Bob Burg and John David Mann, is a fantastic book whose 4th law of stratospheric success talks about the sheer velocity that authenticity brings to every aspect of our lives. That authentic way of being is obvious when it is there and deadly to relationships when it is faked.

CHAPTER ELEVEN- Creativity Barriers, #5 Is Involving Customers

---

"The customer's perception is your reality."

Kate Zabriskie

Get a drink, maybe a snack, and visit the restroom before getting comfortable because this will be the longest chapter in the book.

The starting point for innovation is always Voice of Customer (VoC). That term has a lot of different meanings, including talking to a customer, conducting interviews, focus groups, surveys, monitoring purchase behaviors, etc. At the core, it is about communicating with people who will use your solution or products.

It is very deliberate when we say that communication is with people, not companies. If you ever had a significant project underway with a customer and the critical person left the company, there is always a setback. Sometimes it stops a project immediately, and other times, it is just a bump. It comes back to the fact that we do business with people that work at a company, not with a company that has people. The human element for transformation in marketplaces is the engine that moves us forward.

It is also intentional that there is always communication with the end-users. End-users are not always the customers that purchase directly from you. There are many business models where this is the case. The most common by far are distributors or when you are a component producer. We strongly advocate that you have the distributors or assembling companies in the conversation, but they are not your primary focus.

Despite companies knowing that VoC is vital in a successful launch, they seem terrible at doing it. Some companies go off of a sound bite from a single customer. Some ask leading questions. Some lamely ask what they like, don't like, and what they would improve on some random issue that fits the supplier's needs. However, the worst ones don't even bother to try asking. They come up with their own perception of the market and just make something to sell.

Some companies do very good VoC. They dig in and talk to a variety of customers that currently use their products, some who stopped using their products, and others that have never used their products at all. They may even ask questions to uncover a need the customer wasn't aware could be addressed. They make it about the customer and find what brings the actual user value. However, they can still fall short.

When they fall short, it's because they stopped involving the customer once they started to actually create and missed an opportunity to truly astound the market in new ways unimaginable. The risk factor of missing the amazing solution for the market significantly drops when you bring multiple end-users into the team so that they are more than outside spectators. You have ready-made testers, individuals with different perspectives on the end-use, customers telling you the exact value to them, and real-time feedback if a challenger in the market happens to bring out something new. As Steve Jobs said,

> *"You can't just ask customers what they want and try to give that to them. By the time you get it built, they'll want something new."*

We use a rule before launching any solution to the market. If we don't have sales orders on the books before we launch, we do not launch. By doing so, we make sure we hit the mark and there is a demand for the solution. Launching is not a cheap process. If you don't have the customers that were involved in the ideation and innovation willing to buy it, why would anyone else?

When it comes to having a conversation with a customer, the individuals responsible and accountable for the design/design goals are often not directly speaking to the customer. Most organizations have the sales staff conduct the conversations. This is a massive mistake because of the filters salespeople may have in addition to their own bias. Of course, salespeople should be involved in the conversation, but they should not be the only group engaged in conducting VoC.

The first step is to get a team together that represents different functions within the organization. Diversity (race, age, sex, experience, department, native tongue, introvert/extrovert, etc.) at this stage is not a buzz phrase; it is essential. The more diverse the team, the better because you will have a team that may hear things you don't. Perhaps they listened to what you did but took it a different way. That is pure gold because it allows for questions that bring other points of view. That inclusion and openness allow for Breakthrough Thinking™ to happen. You know you have the diversity of thought from your team reaching a customer if they say, *"Oh, I didn't mean it that way, but that might work if we..."*

The team more than likely should have one member from sales, product management, R&D, manufacturing, regulatory, and customer service or marketing. The first three are essential, with the last three being highly recommended. This team may ultimately not be the ones that carry the project through. They may also not be the people who have potential relationships with the customer. That is OK. It may actually be an advantage. Naïve Experts, which we will discuss later, can be the best people on this team when exploring with customers because they can authentically ask 'why.'

At this point, you may be asking yourself what you would be talking to a customer about because you don't have an idea. That would be a fantastic question if you were trying to come up with a concept and then force it to the market. You are coming up with innovations to revolutionize the market, not iterations. That means you must start with a conversation with key users to find value.

For our discussion, it is assumed that upper management has clearly identified a customer segment and geographic region as their top priority as outlined in "The Five Letter F Word™." Warning, it is a trap to try to develop something globally. There are far too many regulatory and cultural differences to do that for something truly innovative. You won't be ignoring other regions, but that is not your primary concern at this stage. You can always iterate for different regions later.

Here are your steps to VoC and coming up with valued challenges to go into ideation.

01. You want to get twelve to sixteen customers to speak with if possible. Determine who in your market chain is an end-user, the person deciding to purchase (this could be in the same company as the end-user, if not the end-user), a middle man, a deliverer, a group dealing with issues with the end-user, etc. For some, this is a shortlist, and for others, it is longer.

02. Set the number of customers you need to talk to for each level in your market chain. Keep in mind that end-users should be no less than 2X any other group, and that should be if you have a long market value chain. The shorter the market chain, the more end-users and decision makers/P&L (Profit & Loss) owners you should be talking to.

For example, let's say you have a distributor, a servicing company for warranties (maybe for your equipment or for the equipment your product uses or touches), a P&L owner, and the end-users who work in the same company as the P&L owner. You should get two distributors, two servicing companies, four to six P&L owners, and four to six end-users.

03. Now that you know how many customers you need to find, you need to break them into a segmentation. First are the current

customers you have a strong relationship with; preferably at least one large customer and one small. Consider customers you have lost in the past; again, preferably at least one large customer and one small. Then consider adding customers you have never sold anything to, preferably at least one large customer and one small. If you can segment distributors by size, then do so. If not, do it by different physical areas in a region. The same would apply to a servicing company.

04. Have someone on the team that is not the salesperson call the customer. Do not email them first; just call them. You most likely will be sent to voicemail. Here's your message.

> *"Hi, this is (caller's name), the (caller's position) with (your company name). I work with (salesperson name), and they thought you would be willing to help educate me about different aspects of (whatever industry the customer is in relevant to the end-user). I would really appreciate it, and we'll keep the call under thirty minutes. I promise we won't be talking about our products or trying to sell you anything whatsoever. I can be reached at (phone number) or (email). Thank you so much for your time."*

This message needs to be authentic! That message may sound simple, and it is, but it is powerful when delivered as a true statement. A warning, do not, under any circumstances, lie to get them on a call about the time commitment or about selling them anything.

It is highly recommended that the naïve expert is the person who makes the call and ultimately leads the discussions. A naïve expert is educated but not an expert in the topic you are exploring. The power of a naïve expert is they uncover a lot of misunderstandings. They must also be empowered to freely ask 'why' and ask questions that might seem very obvious to others. A great example of a naïve expert is typically a college intern. However, in our example, if the product manager only has what they can find on the internet about a market and its processes, they can be the naïve expert. Also, if you are making an internal process improvement, a naïve expert could simply be someone from a different department.

05. After a week, try a phone call again if you haven't gotten a response. Here is your message.

> *"Hi, this is (caller's name), the (caller's position) with (your company name). I reached out last week, and I know you are really busy. I am trying to get an expert's understanding about (what the industry the customer is in relevant to the end-user). Again, I am not in sales, and I will keep the call to thirty minutes. I'll make myself available for whatever time works best for you and move my schedule around. I can be reached at (phone number) or (email). Thank you so much for your time, and I look forward to speaking with you soon."*

06. Unless it is imperative to have this customer participate, do not reach out a third time. It can put a bad taste in their mouth,

and they may not be as forthcoming. Move on to an alternative. If you need them, get the person in your company with the relationship, usually the salesperson, to set up the meeting.

07. When you set up the meetings, make sure you have the call leader and at least two other team members that can make every call. Keep in mind that your organization has said this is a priority. People must adjust their schedules to address the priority; otherwise, things will become stagnant. Other team members should join as many calls as possible.

08. Get the team into a room or on a call for thirty minutes before the first meeting ever takes place with any customer. Do a ten-minute energizer to get people laughing and their brains engaged. Take the rest of the time to have each person write out ten to fifteen open-ended questions that allow for discovery about the customer.

The questions are to be designed so you are having a conversation where the customer is doing most of the speaking. Remember, you are not trying to sell them anything; only to learn about the what, how, and why of what they do.

Below is a list of several starters that can get you where you need to be with a bit of customization. You will need to write it out to be relevant. Don't make the mistake of just using what is provided below.

A. What do you do at your company?
B. What does your day look like on a good day?
C. What does your day look like on a bad day?

D. What is the single best thing about your job?
E. What regulatory pressures do you face?
F. What do you measure to know you are on track?
G. What does your boss measure to know you are on track?
H. What do you measure to know your employees are on track?
I. If everyone hits their metrics, do you get bonuses?
J. When was the last time you missed a metric, and what did you do about it?
K. How does XXX work?
L. Is the XXX a specialty of your company, or is that how the industry does it?
M. Do all your locations do it the same way?
N. Has anything happened recently that forced you to do something new?
O. How often do you get to communicate with other departments?
P. If you look ten years into the future, and there was someone else in your position, what challenges do you think they would have?
Q. How does your company prioritize (new projects, what is made, when to expand, etc.)?
R. Why do you think your customers purchase from you?
S. Why do you think some customers don't purchase from you?
T. What do you think of our company?

When everyone is done, compile them and remove anything redundant. This is a critical step that many skip; please do not do that. Some ignore it because you won't actually have the list before you during any of the calls.

Do not send the questions to who you are going to speak with, even if they ask for them. If you do, they are going to have prepared answers that kill conversations.

09. Ten minutes before each call, read the list of questions and then set them out of view. The point of creating the questions is to get the team primed and thinking about the customer and not about your organization. Writing these questions out causes a shift in your thinking that is needed for discovery. It allows the conversation to flow more easily because you thought about it before.

Question and answer sessions have a rigidness that comes through in calls. During the calls, the questions are not in front of you or the team because it will turn into a question and answer session. It will stop being a fluid and open conversation. You want the customer to take you where they need you to be. There is an art to doing this effectively because you will not come close to answering even half of the prepared questions.

Again, getting a 123 answer to XYZ question isn't the goal. The goal is to discover what the customer's experience is like each day. When this is done correctly, you will likely learn more things that they need than you can ever work on. By having the other members of the team on the call, who will hear, or interpret, what is said potentially differently than you, you create a wealth of hearing. When you have that vast amount of diverse hearing captured, you automatically enrich your potential starting point for innovation.

The leader needs to start the call with a fast introduction of themselves and their teammates. This should take no more than thirty seconds, and it should not be people introducing themselves. You only have thirty minutes, and this discussion is about them.

10. As soon as you are done introducing the team, start the conversation with:

> *"(customer name) thank you again for taking the time today, and we are absolutely going to keep this to thirty minutes. I really want to understand what you do beyond your title. Can you give me a little insight into you and your world at (name of customer company), please?"*

Usually, they will tell you their title, how long they have been there, and maybe something about the company or their general job. This is perfectly fine. It will give you a foundation for building the conversation. Identifying those building blocks to engage the customer and almost make it personal comes with practice.

11. Once you have something like their title, move on to indicating what you are doing authentically.

> *"Thanks for that introduction. Like I said (in your latest form of communication), we want to learn from you about what you do, how you do it, and more importantly, why. I want to be honest with you; we are looking at the space you are*

in and feel there is an opportunity to improve people's lives. Our method is to understand the industry, so in full disclosure, we will be talking to a few others too. We won't share your name or anything like that or what you say directly about your company. But what we would like to do, with your permission, is talk to you again after talking with others. In the follow-up call, we want to share what we heard from you to ensure we got it right. Then, again without disclosing anyone's identities or secrets, tell you what others said that differed from what you said. We would then appreciate your perspective on the differences."

Do not say anything until they do. The customer may need time to process, and that is OK.

Only once in our history of this method has a group declined to go further. If that happens, say, "*OK, can we still learn about you today*?" If they say no, thank them for their time and leave the call. If they say yes, continue to step twelve.

Occasionally, at this point, they ask about an NDA (Non-Disclosure Agreement). Tell them if they want that, and it's not in place, you can work to get that done. However, reassure them that you aren't asking about secrets or things like that. You are only interested in understanding what they can freely disclose about what they do.

12. At this point, the art of conversation comes to life. Your task is to keep them talking with as little dead time as possible. This is why you prepared the questions ahead of time, so that you can move swiftly forward based on what they are saying. Your

jumping-off point would be about them and their job if they didn't give a detailed description. From here, you can say,

> *"I want to keep to our time, so we aren't wasting yours. You said that you are a (job title as they said it). What does that look like on a good day for you, like, what's in your daily activities?"*

You are getting them to talk about themselves as an individual first and extracting the positive. When we express genuine interest in the positive aspects of people's lives, they talk more freely.

13. Keep the conversation moving based on what they are saying. You will find that the prepared questions will naturally modify themselves as you explore areas that they lead you to. Don't overthink it.

14. When there are about five minutes left, and without interrupting them, say,

> *"Listen, we have a few minutes left, and I want to thank you for this incredible conversation. Do you (customer name) have any questions for someone on our team?"*

Pause for a moment and let them think if they have a question. If they do, then answer it honestly and quickly. If someone on your team has a question, it should be to clarify something the customer said. It shouldn't introduce a new topic.

Our goal is to finish the call on time. If the customer continues to talk and ask questions past the time frame, let them.

15. Finish the call with,

> *"Again, we appreciate your time. I will reach out in the next two to three weeks and set up our follow-up call. I think you will get a lot of value from that sharing, especially where we heard differences. We will also send out our notes from today's call once we compile them. How does that sound?"*

Repeat this process with each customer on your list. You are going to find the conversations can be wildly different. There may be other questions asked to each customer with little overlap. That is perfectly fine, but you will discover common aspects from like groups in your market chain.

You might be wondering at this point why we haven't talked about any ideas yet. The truth is we haven't been ready to, but we are getting close. We need to complete the second call with the customer because it sets the essential framework for the ideation.

16. As a team, set aside three hours to go through all the information you gathered. Start with an energizer and take a five-minute break after reviewing every third customer. You will need a giant whiteboard for this, but if you don't have one, try MIRO. It is an endless digital whiteboard, and you can sign up for free. It has a lot of valuable tools that are reasonably easy to use. You can also use excel, but it isn't recommended because you are capturing whole thoughts, not just a word or two.

Start by setting up a simple layout. In row one, set up the company name, name of the person interviewed, date of the interview, conflict 1, conflict 2, conflict 3, conflict 4, conflict 5. Populate the first three columns for all of the customers.

17. Take the first customer on the list. Each person should scan their notes from the call and find what they think an area to explore could be. Maybe the customer is having trouble hiring qualified people due to their location or just people in general. Maybe they need new equipment to operate more effectively. Perhaps their internal documentation processes are so cumbersome that they can't change because no one wants to process the paperwork.

Find at least three, but no more than five conflicts, per customer. The team might have different interpretations of what was and wasn't said. That is great because it forces a conversation that can change the importance of that potential conflict. Your team discussion should also flush out what the most important ones are if there are more than five for a given customer.

Please do not fall into the trap of thinking which one might be the most significant dollar value opportunity for you as being the customer's biggest struggle. If you do, you will find, as outlined in "The Go-Giver," that you are going to miss out on the first law; The Law of Value. Simply put, this says you should always give more in value than you receive in payment. You are not worried about payment yet, so only focus on their needs and what is valuable to them. Remember that money is always an outcome of goals that hold value, not the other way around.

Repeat this process for each customer independent of all the others. There will likely be a lot of overlap, which could signal the most significant market value, but that is a premature assumption.

It is crucial to capture the differences when and if they come up. It is these differences your customers want to hear about on the next call and what you indicated you would provide them.

18. After everything is captured, write out one sheet per customer and have it in front of you on the next call. Each team member (assuming you have six) can do two (assuming twelve customers). The sheet per customer should be the three to five things you heard that you need to make sure you understood. Also, it contains anything a different customer said that could apply to them.

Be mindful if a difference is likely irrelevant from one customer to another. An example would be a distributor with logistics challenges due to time zones. An end-user that only works from one location won't care about the logistics issue. Generally speaking, you shouldn't exclude the differences unless it is evident, like in that example. The customer may have never thought about that difference, and it might be an issue or solution for them.

19. Within the time frame you established, get the call scheduled and keep it to thirty minutes. Send them the complied notes specific to their call in the meeting invitation. Don't send them any

differences or information from the other companies. Those will be your areas of discovery with them during the conversation.

20. Start the call with,

> *"Thank you again for taking the time to talk with us. Today we want to make sure we understood some things. We also want to get your thoughts about what others said and why you think they might be doing that. I also wanted to let you know after we talk to others, we will do some fun ideation sessions on Breakthrough Thinking™. It is how we start our innovation process here at (your company name). You are invited to participate if you want. We'd love to show you how we are working on ways to transform the industry. If you aren't able to participate, that's OK. We'd like to have one more call to describe what we think could impact your job and why we think it brings you value. It would be a final check to see if we got it right before we start doing work on it."*

Keep the call energetic and humorous. You can have your other team members check on aspects you are verifying if you want. Just make sure after any of you explain the conflict back to the customer, you ask a question similar to,

> *"Did we get that right?"*

When you get to the differences, make sure you are succinct so the customer can consider the points and talk about them. If they make a comment similar to *"we haven't ever done that,"* then you are about to discover value. They will tell you why they haven't, which is very insightful. If they don't say why then they are likely to start talking through why that is of interest. Now you have the makings of the most prominent ideas to ideate around in addition to the things that were common across multiple customers.

Try to keep the call to thirty minutes, but if it runs over because the customer is actively engaged, you go until they have to stop. They won't mind. When you do end the call, ask two questions. First, ask if they got value from the meeting. They will almost always say yes. Then ask if they think they might want to participate in the ideation or if you can check back with them later when you have some dates and times when the team will be ideating.

Most won't participate because of the time commitment and trying to align schedules. Plus, the customer would have to travel or attend over a webcam service that isn't as effective as in person. If they do want to attend, it is a fantastic ideation session waiting to happen.

21. As a team, you'll have to decide what concepts a larger group will ideate around. This is not the ideation itself but a slight form of reduction. This reduction is specific to giving guidance only, so the larger team has a starting point.

A. The team members look at the list of conflicts they feel your company can act on related to your company's overall established strategy to find challenges. These challenges will be a combination of the overlapping conflicts and any differences that seemed to spark interest from multiple customers.

B. Each participant selects the top five they feel are the most feasible and most attractive. Feasibility involves having, or having easy access to, potential expertise, equipment, raw materials, staff, and space. Attractiveness consists of getting to scales of economy, potential intellectual property, compliance with regulations, possible market size, and potential profitability.

This is a best guess. If someone is struggling to do this, then assign a value for feasibility, then attractiveness, from one to ten for each challenge on the list. Ten is the best and one the worst. Once they have assigned values for each listed item on the two parameters, then multiply them together per item on the list. The top five win. If there is a tie, then default to one that overlaps with other groups identified in the initial round of calls.

Your best guide at this point is going to be intuition. Intuition is the feeling in your stomach that says yes or no based on loosely connected experiences that haven't formed neuroplasticity. This is different than instinct, which is just done naturally, like suckling. Intuition is often hard to say why you feel that way, but you just do. In this situation, trust your gut because you will have time to verify later.

A small warning about intuition. It is very quickly skewed as verification within your thinking due to personal bias. If you feel it in your gut, it is your responsibility to find data to verify your thoughts. You also must actively search for information to prove you wrong; otherwise, you will fall victim to a biased belief. We call it testing the null.

C. Once each participant writes out their selections on a sticky note, place them on a wall. There's bound to be overlap. The top three items that get sticky notes are the challenges for the larger group to explore with an ideation session. If there is a tie in the group, do a simple vote. If this is your first time doing this, only take the top challenge to the next stage.

22. This last step is vital to how the ideation will go. You need to use no more than three sentences to describe each of the challenges you just selected. The sentences should be as simple as possible and include the 'why' of importance. The 'why' gives guidance without real reduction. Each member of the smaller team needs to attempt to write this out independently.

Here is an example:

"Our customers are ideally located in more rural areas who do not have population centers within fifty miles. Most of the staff they can get do not have a high school level of education, and some do not speak English well. They need a solution that allows them to expand without adding staff, complexity, or more paperwork for their industry."

This includes the customer's conflicts and gives a sense of the need. It doesn't suggest any solutions or restrictions to solving the issue. The goal is to take this into ideation, where you seek bold and audacious ideas. If you position it with heavy restrictions too early, you will miss out on potential ideas and potential team members wanting to work on the project. There is plenty of time to reduce the concept later.

CHAPTER TWELVE- Creativity Barriers, #6 Is Stuck, Diluted & Killed Ideas

---

"You can't kill ideas. But you sure can shoot the people who have them."

G. Gordon Liddy

Before we get into an ideation session and how to conduct it, we need to understand a few more topics. First is the last major barrier to creativity. When ideas become stuck, diluted, or killed before their time, you lose momentum and the eagerness of the team to work on them. They become a reflection of the company's inability to prioritize and move forward. People will lose faith.

If someone in an organization has an idea, and it takes months before it's reviewed, the person who submitted it would potentially feel ignored, leading to resentment. That can also hold true if they're simply never informed of what happened to the idea, even if it was reviewed. There is far more damage done by not following up with an idea originator than there is by killing it too quickly.

The chances are that person will not submit other ideas again. There is a likelihood they will disengage from their job and tell others about it. A cascade effect can occur where individuals view upper management as disingenuous, only saying they want everyone to participate because it sounds like good public relations. The employees distrust anything upper management says because the employees feel they are not valued.

The dilution of ideas is the arch-enemy of Breakthrough Thinking™. The practice of dilution is when you try to make an idea for everyone and please no one. The final products derived from diluted ideas are slightly improved versions to your market challengers solutions (A.K.A. me too solutions). The 'new' product only holds value for someone new to the market. It is estimated that unless the value of something new isn't at least 20% greater than what they are doing now, they will not change unless forced to by regulation.

The resistance to change is, in part, a response to the unknown. David Gleicher proposed a formula that was later modified by Kathie Dannemiller (C=D*V*F>R). This formula says for Change to occur, the Dissatisfaction with how things are presently is multiplied by the Vision of what is possible and then multiplied by the First tangible steps taken towards the vision. That outcome must be greater than the Resistance. If any of the factors of dissatisfaction, vision, or the first step are zero, you can't overcome resistance.

The reluctance to change is so great, according to Dr. Robert Cialdini in his book "Influence, The Psychology of Persuasion," that people will go out of their way to create avoidance to change. They can have strategies such as ignoring the source of the change, contesting data they would usually accept, citing regulatory or corporate policies that aren't real, deferring responsibility, claiming contractual obligations, or misrepresenting data to decision-makers.

The killing of an idea before its time creates limits too soon. Most ideas get killed before their time because they feel too big or too outside of the normal scope of projects. Often that is exactly what is needed to move the needle forward. Since companies kill the big, crazy, and wild ideas, many seasoned idea presenters dilute them to fit a model for internal approval that fails external approval. Those types of ideas are outside of the normal comfort zone. However, if you have communicated with your customers to seek their value needs, ignore the requirements of your company until later. Being outside of the comfort zone is what allows you to transform the market.

A lot of companies struggle with when to stop or kill an idea. The short answer is when it is time to commit resources to form a project, then you eliminate most of the pre-project ideas. This process of elimination is known as the money gate, or, in traditional Stage & Gate, Gate 3. In "How Are Kids Innovating Faster Than You?™," we outline this in more detail. However, during our ideation process in Breakthrough Thinking™, there is a natural elimination of ideas that doesn't result in demoralizing teammates.

Reducing the number of ideas that become projects with substantial resources behind them is important. It forces focus on the exceptional ideas, not just the good ones. A way to think about this can be seen in public transportation. Ideas are like a bus; there is always another one around the corner. It is your job to catch the right bus that takes you on the journey you choose.

SECTION FOUR: UNLEARNING FEAR

CHAPTER THIRTEEN- Best Ideas And Neuroplasticity

"Let your ideas and thoughts give you inspiration. All creativity comes from your imagination – you first imagine, and then you create."

Catherine Pulsifer

Where are you when you get your best ideas that inspire you and get you excited? For more than 99.9999% of us, it is not at work. In fact, during our live Ideation Emporium of Creativity™ sessions, we have never gotten that answer. The common answers are when we exercise, cut the grass, take a shower, drive, cook, or maybe just as we fall asleep or wake up.

Why is that? It comes down to how our mind operates and what we are doing. It is influenced by the brain waves being manifested in combination with the physical environment. Those two factors invoke our neuroplasticity. This section will bring awareness to allow you to employ Breakthrough Thinking™ whenever and where ever you want to deliberately.

Take a moment and try to create a mental image for yourself of the perfect day outside. Perhaps you are just arriving at an open beach area near a lake. The lake is fed by a river flowing from a magnificent mountain in the background. Outside the beach area is a forest with different trails for walking and biking. The area isn't crowded, but there are others enjoying the scenic location by doing everything from swimming to yoga to just kicking back in the sun.

Past experiences like this bring a smile to your face without you being aware of it. You feel yourself relax and maybe even reflect that this is what life is all about. You find it easier to laugh and are more open to suggestions about what others would like to do. It is a great day.

Maybe an hour later, you are swimming, and an idea strikes you. The idea is one that makes you pause for just a moment. It is exciting and feels like an opportunity to create. In that moment, you are in what we will call Aspiring Alpha. Your brain is relaxed, and your neuroplasticity, which we will define in a moment, is engulfed with latent positive memories of situations like this.

Aspiring Alpha is that mental space where your mind can crack open the door to the subconscious. Our brains let go of being in Basic Beta, where we spend most of our day. When we tap into Aspiring Alpha, we are typically doing something that doesn't require direct attention, like being in the shower. The mind can wander just a little because we are not in an environment that requires our expertise or conscious focus.

When our expertise is required, we are in Basic Beta. Basic Beta is not an auto-pilot because as we move from task to task, we need to comprehend what is happening. The mind is focused on what we do every day. Being in Basic Beta is like being in a deep river rushing down a mountain that we can almost navigate with our eyes closed. We can smoothly maneuver within our rivers because of our neuroplasticity.

A simple way to think of neuroplasticity is to think of it as a collection of all emotions, touches, smells, images, thoughts, and experiences we have ever had. The more we experience things over and over, the more ingrained it becomes into our brain.

As an infant, our neuroplasticity is just forming because we can't do much. Toddlers seem to be like sponges, just absorbing everything around them because they are at the basics of building neuroplasticity. As we age, our channels continue to get deeper as we revisit them more often.

If you have ever been in a deep river with a swift current, you know how challenging it is to escape. If you relax, find your single target, and work towards that, you can do it.

Neuroplasticity, which is analogous to the river, can be overcome the same way in regards to our experience. If you have high levels of comfort in your expertise, you have to work to break out of it to find growth because you are moving the river channels of your brain. The deeper the track, the more complex the struggle is to change it. This is part of the reason change is so difficult.

Is there a song you haven't heard in years? Despite the time that has passed, you can hit every word on the right beat like it was yesterday, and you had it on repeat in your bedroom as a teenager. It is embedded by your neuroplasticity. Our younger experiences are so profound that those with forms of dementia can often recall things sixty years ago with vivid descriptions while perhaps not being able to recognize their grandchildren or even children.

There is an academic argument that neuroplasticity can change for all aspects of life given enough reinforcement and time. Unfortunately, that argument falters with those having undergone trauma, especially at an early age. Fear, and the chemical release that comes with it, seems to be the fastest and deepest builder of neuroplasticity. That makes some change so overwhelming that an individual might not be able to get out of the river, especially alone.

A simple experiment can demonstrate how we can change our neuroplasticity. In a comfortable seated position, take your arms above your head and have your palms face each other. Without thinking, bring your arms down and cross them over your chest. If you repeat that but cross them the other way, it will most likely feel foreign. However, if you raise your arms and deliberately cross them the uncomfortable way forty times in a row, three times a day, for twenty days, you will change your neuroplasticity to have that feel more natural than the way it was before you started the experiment.

When we are in Basic Beta, we can struggle to get out-of-the-box thinking started. Just like when we cross our arms, Basic Beta is comfortable. We want to be in Basic Beta most of the time, especially at work. After all, that is why you are in the position you are in, for the expertise you possess.

When we are trying to ideate and come up with new, sometimes crazy, ways of doing things, we have to escape Basic Beta. The fastest way to do that is through laughter. Laughter actives our brains in different ways, releasing catecholamines and repressing cortisol. When we are having fun or relaxing, our brains can crack that door open to get to Aspiring Alpha, just like in our swimming example above. Laughter can do this too.

What is the best present you can give a five-year-old? A box. Why? Because they can make it anything they want. The bigger the box, the better. It is an unlocked, unbound world where they can freely create. That type of child-like thinking brings out the best ideas in Breakthrough Thinking™.

Earlier, we mentioned the need to think child-like without acting childish. We also noted the incredible amount of laughter children perform compared to adults. They have that ability to just imagine and create. It is an ability you still have, you just need to unlock it so you can get out of your river of neuroplasticity. That is the essence of, the importance of, having fun and laughing at work. It unlocks you. Aspiring Alpha, through laughter, is the key.

There are two other places where some of us get our best ideas. The hardest to capture is Dang It Delta. Have you ever been asleep and suddenly woke up knowing you just solved world peace? As you start to move to write it down, the idea vanishes in an instant making you say, "Dang it," or perhaps some other choice words. In Dang It Delta, we are fully engaged in the subconscious. Returning to the conscious mind simply breaks those delicate threads of thoughts as we return to neuroplasticity.

Some people get their best ideas just as they're about to fall asleep or just as they're waking up at a regular time. That is called Tremendous Theta. The door to the subconscious is wide open, but we haven't wholly gone through it yet. An example of someone like this is Thomas Edison. He coined the term "when the penny drops," which often refers to when an inspiring idea strikes us. He placed a tin pan on the floor under his knees when sitting in a chair. He then would place a penny between his knees and keep them together. As he would start to fall asleep, his legs would relax, and the penny struck the tin plate with a clang that woke him up. He then proceeded to write down anything he was thinking of. He certainly had some Breakthrough Thinking™.

CHAPTER FOURTEEN- F.E.A.R. Of F.A.I.L.ing

"Being afraid and fear are not the same thing. Being afraid is about surviving a physical threat in front of you. Fear is about your brain having a presumption about something it made up."

We have touched on several aspects of fear already. Fear is part of our most basic response to danger. When a physical injury to ourselves, loved ones, or property can occur, we react. It is often referred to as Fight-or-Flight response. That response has kept us and many other species alive for millions of years. This shared response across multiple species is what we will refer to as being afraid. Being afraid or scared is just a tiny part of fear in humans.

The remainder of fear in humans is not real. It is something we make up in our heads. It often comes from the unknown. It is then magnified by others within our social groups' expression in regard to the novel subject. Other terms that we use for fear in our heads, or other groups, can be worry, stress, doubt, frustration, anxiety, nervousness, panic, uncertainty, and failure.

In many areas of our lives, especially in the corporate setting, fear is the elephant in the room. People fear looking bad or saying something against someone else's thoughts. Then we fear failure and what could happen. Far too often, it is because trust is not established, egos get in the way, society puts us into ranks. Fear wins as it reinforces neuroplasticity, as discussed.

None of it is real. The individual is not stronger, more intelligent, creative, or more inspiring than the group. Challenging ideas should be welcome and encouraged; otherwise, you will never escape your own neuroplasticity, and neither will the ensemble. Letting go of the need to be right in front of others allows you to grow while fostering relationships with others. Doing so allows you to take fear head-on as a community. In this way, you reinforce our need to belong while making it part of the culture to vanquish unrealistic expectations of perfection.

The best way to think about F.E.A.R. is as an acronym. False Evidence Appearing Real. In that light, you should also consider the word F.A.I.L. to mean First Attempt In Learning. Remember that Thomas Edison said out of the thousands of attempts they had at creating the light bulb, none of them were failures. They were ways that they learned not to do it.

Consider a baby just learning to walk. When they stand up and take those first few steps and fall down, what do you do? Do you start correcting them or scolding them for failing in the attempt? No, of course not. You support them and encourage them to try again. You praise the few steps that they got correct.

If you aren't helping people in your organization to try new things in this manner, why would they ever leave their comfort zones? If they can't leave their comfort zones, your company will never achieve Breakthrough Thinking™ which creates wide-open markets that others aren't in. Thomas Edison wouldn't have revolutionized the lighting market if he had only sought to improve the candle.

CHAPTER FIFTEEN- Kids -Vs.- Adults

"Adults are obsolete children."

Dr. Seuss

If we were to list the characteristics of kids, it probably would be:

01. Unfiltered
02. Imaginative
03. Honest
04. Emotional
05. Happy
06. Loud
07. Full of Energy
08. Easily Distracted
09. Ask Why All The Time
10. Intrusive
11. Messy
12. Fast Learners
13. Explorative
14. Forgetful
15. Collaborative

If we asked for a list of characteristics of adults, it might look like this:

01. Conservative
02. Forward-thinking
03. Structured
04. Tired
05. Tactful
06. Responsible
07. Gets Things Done
08. Practical
09. Focused
10. Quiet

The lists certainly are not all-inclusive, but they do allow for contrast. If we were to utilize a single word to describe kids, it would be expansive, while for adults, we would use reductive. Neither is good or bad nor right or wrong. We do actually need both in the world to operate.

As discussed before in The Long & The Short, we need both sides to make the whole. When we ideate, we have to incorporate expansive and reductive points of view to drive solutions to the market. Just like in The Long & The Short, these have to happen in tandem, but not simultaneously.

We need to spend about 20% of our time expanding ideas and 80% reducing them. That may sound backward, but expansion is incredibly fast once you unlock your creativity. The struggle is reducing ideas while not letting them become stagnant, diluted, or killed before their time. The reductive part of

Breakthrough Thinking™ requires more time, focus, and incredibly intense creativity.

Within Breakthrough Thinking™, we always begin by expanding with child-like thinking on the customer's conflicts during ideation. We bring together a team of twelve to twenty individuals. We then get them energized about conducting an idea-building session using various tools that we will describe. Following that, we will do a speedy reductive selection of the ideas to move forward with the team.

SECTION FIVE: CREATIVE IDEATION

CHAPTER SIXTEEN- Building Up Ideas

"Bring a brick, not a cathedral."

Building up ideas is a simple concept, but that doesn't mean it's always easy. The challenges will come around not taking shortcuts, staying focused, and building an ensemble to keep each other moving forward.

This is slightly different than ideation. Ideation can be expansive and reductive because ideation requires creativity. As we just discussed, reduction requires intense creativity. Building up of ideas is mostly expansive.

Our starting point for building up ideas began with the VoC conflicts the smaller team has collected. The next step is to bring it into a larger group that has diversity and who will be expansive in their thinking. You will set the stage for ideation by indicating that it is a 'Yes, And' session, a 'Plus-ing' session, or an 'Unbound In Reality' session. These are all the same thing, just different ways to express them depending on your culture or who may be participating.

Start with an energizer appropriate for expansion before anything else. Get everyone smiling and laughing. You need to have everyone in Aspiring Alpha to get the building up of ideas to flow freely. By starting with the energizer, you are also engaging playfulness and team building.

The next things you want to establish for the building up session are the goals and the rules for the day.

GOALS

01. Generate a lot of ideas that are simple to understand very fast. You are going for quantity, not quality.

02. Everyone participates. Introverts, extroverts, all positions, everyone.

03. Have fun in the session and avoid thinking any idea has to be right.

RULES

01. This is not a reduction session. Phrases or thoughts to avoid

A. That will never work.
B. No, because.
C. Yes, or.
D. I don't know.
E. This isn't right.

02. Bring bricks, not cathedrals. No one person should be dominating the idea so much that others can't add to it. If you already have a cathedral, there isn't much that can be added. If everyone has a bunch of bricks that they can bring, a cathedral can be built together. The cathedral of an idea

is built together through teamwork and everyone bringing their idea bricks.

03. The only person in a group of four that should ask questions is the Capturer. The Capturer records the ideas on a Capture Sheet, which we will describe soon. They are responsible for asking questions that add to the concept, such as: what is the container like, how are we distributing this, what will it cost, etc. Nothing has to be correct; it just needs to be captured in the confines of five to six sentences.

The Capturer, a rotating position, is also responsible for ensuring that whatever the first suggestion given is, that one is built upon with 'Yes, And.' They need to get everyone offering ideas by not allowing someone who just gave a brick to an idea to continue adding bricks until others have. Here is an example using cupcakes.

> Person A: *The cupcake will have amazing sugar free icing,*
> Person B: *Yes, and that icing will be all-natural,*
> Person C: *Yes, and those all-natural flavors will also have all your daily vitamins,*
> Person B*: Oh, and they are done by age needs for the vitamins,*
> Person A: *Yes, and they are gluten-free,*
> Person C: *Yes, and they are nut-free,*

Person B: *Yes, and they can be ordered online for pickup in one hour.*
Capturer: *What about pricing? And are we able to deliver?*
Person A: *Let's do $2.00 per cupcake with the 12th free if they buy 11.*
Person B: *I like that with $0.25 per cupcake for specialty upgrades.*
Person A: *Oh, yes. We can also waive that fee for first-time orders on their next order if they leave a review.*
Person C: *I like that. Delivery can be Uber Eats or Grubhub, but on larger orders, we can do delivery for $15.*
Capturer: *OK. That is definitely enough of an idea. Let me see if I can write this pitch out.*

"The all-natural cupcakes that you can customize online and pick up in an hour. These sugar, gluten, and nut-free cupcakes come supercharged with your special daily vitamin needs based on your age in any flavors you like. Starting at $2 per cupcake, and $0.25 per cupcake for each customization, you can now enjoy delectable treats anytime you like with delivery with your favorite food ordering service. When you order 11, you get the 12th free. New customers can get the $0.25 per cupcake upcharge fee waived on their next order after they leave a review on their first order."

Person C: *That is good, and I agree the delivery fee doesn't have to be in that. It is too much detail on how it works. Nice summary.*
Capturer: *OK, what is a title? Remember, make it something they will be curious about, like click-bait.*
Person B: *Granny's Secret Cupcakes*
Person A: *Oh, Granny's Secret Cupcakes: devilishly tasty, heavenly good for you.*
Person C: *OK, and we can draw a box with cupcakes with horns and others with halos.*

Please denote that the Capturer did not offer out any ideas. They did, however, take what they heard and form the pitch. That is how they contribute. Also, remember that everyone takes a turn as the Capturer.

Following setting the expectations for the session, briefly introduce the customer conflicts you and the smaller team came up with. Read out the three sentences per conflict. Do not add to them or give further explanations. You do not want to restrict thinking.

Hand out copies to each four-person group so they can refer to them. Then let them begin. They are free to utilize any expansion tools they like as long as they capture the ideas, keep the concepts simple, and generate a lot of them.

Remind the team one last time that 'Yes, And' is about adding to what another person has said. It is about affirming that you heard them, then showing how you want to add to what they said, not change it. Be wary of 'Yes, And' that is 'Yes, Or.' 'Yes, Or' is someone trying to impose their idea over someone else's, not to add to it. Have them hold the 'Yes, Or' for a new Capture Sheet.

CHAPTER SEVENTEEN- The Tools

"If the only tool you have is a hammer, you tend to see every problem as a nail."

Abraham Maslow

You don't have to use all of these tools. However, if you are stuck, try one that you are less comfortable using. Some of the ideas that will come out will be crazy, silly, outrageous, audacious, and laughable. The key to these tools is remembering that the outcome doesn't have to be right to be useful. The usefulness may come by inspiring another idea that becomes the perfect fit.

01. **THE CAPTURE SHEET**- This is a simple way to capture an idea. It is a piece of paper divided into three sections and is easy to recreate anywhere. First, but usually done last, is a title. Try to make your title relevant, but almost like click-bait. It should make the reader curious. Second, on the left side of the paper is a space for a drawing. A drawing can genuinely make or break the idea. Third, on the right, is a space to describe the project in five to six complete sentences. Using

whole sentences is vital as it forces you to keep the ideas simple. You should be able to hand the sheet to a person not in the room, and they get the basics of the idea without you talking. That means not a lot of unneeded detail. Many times, Capture Sheets will overlap with other Capture Sheets from the same session. That is probably a theme for exploring during reduction. We have a template for download at www.IdeationEmporium.com; the password is "innovate."

02. **BRING A BRICK, NOT A CATHEDRAL** (Yes, And)- You start with the first idea someone says. Then the next person says *"Yes, And..."*. They actively listen to what the other person said, letting go of any judgment about what was said and adding to it. You don't have to worry about being right, just adding to what your teammate said.

03. **COSPLAY**- Pick any person. They can be alive or dead. They can be real or fictional. They should be reasonably far from what you are trying to solve (e.g., don't use Albert Einstein to solve a physics problem). Try to put yourself in their shoes and think like how they would approach the problem. Now, have someone interview you and answer some questions. Have some fun with it and even try out their voice. If you are a real enthusiast, go full cosplay in a costume too.

04. **ALREADY INVENTED ELSEWHERE**- Describe the problem as you would to a nine-year-old. They know enough of the world

to function but not as much as a twelve-year-old. This gets the problem to its simplest form. Take that simple statement and look over the world to see where it might be already solved. Apply aspects of how others solved similar issues. There might not be a perfect fit, but components might jump out. Capture those on a Capture Sheet.

05. **REFRAMING/HOW ELSE**- Take portions of the problem and explain it differently. More importantly, use empowering, different, outlandish, upbeat, and fun words. Causing a shift in the language can lead to fantastic breakthroughs! An example is how Disney doesn't have employees; they have on-stage and off-stage cast members. Another would be instead of labeling someone as a receptionist, they are a director of first impressions.

06. **ATTRIBUTES**- Pick a random object, not a pen, pencil, computer, desk, or chair. Google is great for random suggestions if you can't seem to find one. Make a list of all the uses and physical attributes of that random object that you can. Now try to apply each use and attribute to your problem. This won't solve the issue, but it will make getting out of your river easier.

07. **BREAK IT**- Think of an area where your intuition feels conflict. Write down every rule that you can think of that allows that conflict to stand as it is today. Now, select one rule

and break it by asking, what if. Just like Netflix asks, what if late fees weren't a thing. You will find that other rules start to break as you do that. Generate ideas and make sure that you are writing them down. Remember, simple is best, as you will see when using the Capture Sheet.

08. **SONGWRITER**- This tool has a specific use. If you are struggling to get to the heart of an issue, write a song. Pick a tune that you know well and follow the pattern of that song. The lyrics must have a familiar rhyming scheme that will force you to include words that make it wrong in the opening part. When you get to the refrain, and this is the most critical part, think about what you want to repeat over and over. Write it out. If you wish, you can do one more verse. Now, as a team, sing it. It will likely be wrong, but we bet it is beneficial. Using the wrong words to describe something makes it easier to find the right ones!

09. **MIND MAPPING**- This can be done with a simple piece of paper or using some free tools readily available online. This tool is useful when doing Voice of Customer to connect abstract ideas. It gives you a visual representation and allows you to understand in real-time where you might have a gap in your listening.

If you are alone and just feeling stuck, you can try some of the above tools. However, here's a list of things you can do privately. You should denote that a physical component is often needed to help the shift to Aspiring Alpha when you are alone. If you feel silly, that is great; it means you are out of your comfort zone.

Things to do if you are mentally stuck-

01. Sing
02. Dance
03. Get into a cold shower
04. Read aloud and be loud
05. Count from 201 backward to zero by sevens
06. Alphabet game- use each letter of the alphabet to name:
    A. Animals
    B. First names
    C. Last names
    D. Cities
    E. Objects

07. Watch a video that makes you laugh every time
08. Eat something very sour or bitter
09. Eat a lot of chocolate FAST
10. Make a holiday present list
11. Do some stretching /yoga
12. Do fifteen jumping jacks, fifteen sit-ups, and fifteen more jumping jacks, followed by thirty seconds of running in place as fast as you can (get those knees up!)
13. Find a quiet area and do some breathing exercises
14. Do some hardcore cleaning
15. Do some line touch basketball sprints

16. Play catch
17. Ride a bike
18. Out loud, call out five things you can see, four things you can touch while touching them, three things you can hear, two things you can smell, and one thing you can taste
19. Anything requiring you to follow a pattern or instructions like baking or knitting
20. Anything that makes your body move
21. Anything that makes you laugh

Things NOT to do when you are stuck-

01. Stay still where you are without breaking the environment
02. Stay engaged in your thought process
03. Act on impulses unless it is an action from above
04. Continue talking to the person you are engaged with
05. Feel guilty

CHAPTER EIGHTEEN- The First Reduction Round

"We are reducing the timeline by reducing the non-value-added wastes."

Taiichi Ohno

Now that you have many ideas, you need to get the very best of your bricks to make your cathedrals during your innovation process. At this point, you are just starting to focus. This is not a business case, so we are not yet worried about feasibility, just the attractiveness of the idea to the ensemble.

Explain the process in full before starting so everyone can see that this will be anonymous. Hang all of your ideas on the wall. Have the group sponsor, the person that called the meeting, go through all of them and read them aloud to everyone. Yes, do this! As ideas are read out, some ideas may naturally start to fall into groups. It is okay to move similar ideas together, but remember, each Capture Sheet is an individual. Assign an individual number to each Capture Sheet.

Now, a tiny bit of math. Count the number of Capture Sheets; this should be easy as you just numbered them. For every seven Capture Sheets, give each participant two sticky notes.

These sticky notes will represent a vote. Each participant now takes a few minutes to walk around and refresh their memories of all of the ideas they just heard. They then write down the number of the Capture Sheet they want to vote for, one vote per sticky note.

Get everyone into a big group and start walking. As people walk randomly around the room, they exchange sticky notes. It is a constant swapping of notes. No one can refuse to trade, and everyone should have two sticky notes, per seven Capture Sheets, at any time. After about sixty seconds, stop. Have people go to the Capture Sheets and put the sticky notes with the corresponding votes next to them.

The team has now reduced the ideas to the tops ones for consideration. Even if someone did not create or even vote for the idea, they are still part of the team. They have contributed to the process and have buy-in as a team member through the creative and playful approach.

There is one more tool, at this stage, that is helpful in reduction if you find you have too many options to choose from still. It's called Flower Power. This tool positions the Capture Sheets that the larger team voted for against the three-sentence criteria you presented during your initial setup, after your first energizer.

Take the top three criteria per conflict you had and arrange them in a circle. Draw a line to the center of the circle for each of the criteria. This is an axis to make your flower on. Divide each axis into three equal parts. Label each section nearest the center as a one, the middle a two, and the farthest a three. Each customer conflict gets a circle.

Write out a sticky note for each idea for each axis. Have the team vote on the strength of the concept as a one, two, or three per axis. Three is the strongest or best. After you have voted each idea on each axis, see which makes the biggest flower. That is the idea that moves to innovation.

CHAPTER NINETEEN- Customer Feedback

"Don't waste customers' time asking them questions unless you are prepared to act on what they say."

Bruce Temkin

To prepare for the next round of calls with customers, you will need to prepare a value pitch. Each participant for the twelve to twenty-person group will need to do one per idea. Later, the smaller six-person team will consolidate them to form the strongest one. Remember, when you list anything, it needs to be what you think the most valuable aspects are first.

Adlibs are never perfect, so you may need to modify this slightly to flow for your situation. You present this as a minor press release or a mention in a news article.

> *(Your company) is pleased to announce (name of the product). This breakthrough solution delivers value to the (name of the industry) by (value one, value two, and value three). It meets or exceeds all of the regulatory requirements in (name the region) and has accreditation with (if accreditation is necessary, name it here). It is ideal*

for anyone who has to (describe what the end-user does in one or two sentences. Be concise).

It works by (describe how it provides value in four to five sentences as you did in the Capture Sheet. When written correctly, you don't have to call out the value again; it is self-evident from what the customer told you).

This product will be available (pick the first reasonable date that you are more than 70% confident in that you will hit).

When the smaller team is consolidating, use as many words as you can directly from the customer's vocabulary. When you use other people's words, they are more open to listening.

Do not send the material to the customer beforehand. When speaking with them, read it precisely as written. Then ask them two questions. On a scale of one to ten, ten being the best, what VALUE does this bring you? On a scale of one to ten, ten being market changing, how DIFFERENT is this for you?

They will more than likely start talking out loud. That is a fantastic sign because they are thinking. Make sure that you get the actual quantification from them. This quantification is important as you move into the innovation process, as we discuss in "How Are Kids Innovating Faster Than You?".

If you get a question about price, answer it honestly. Say you do not know yet because you don't. Stop talking at that point about price or costs. Don't try to justify it. They may say things similar to, "*Okay, well, it needs to be reasonably close to X price because…*". Take note of that. That is the paid amount for the value they get now.

When you are focused and deliver on the value they need, they will pay more. At least you now have a benchmark of the lowest price you'd have to consider selling for, which is helpful in business case projections. However, you also now have a quantification to derive the value difference that your new solution brings.

At this point, you would begin your innovation process. Keep in mind you are innovating with this solution, not iterating. Stage & Gate processes aren't bad, but they are cumbersome and designed for iterating. Whatever you do, move it forward with focus and deliberate actions.

Make sure that you schedule regular updates with the customers you have involved up to this point. They would appreciate updates because they would like to hear about real developments. Also, they will tell you if you are on track or not. Plus, scheduling those updates at the start of the project will help you and the team maintain momentum. There is nothing like saying, *"We need to update multiple customers,"* to motivate upper management in helping to move things along.

CHAPTER TWENTY- Time For Transformation

"The era we live in are dreams come true. Will the next era be made of your dreams?"

Fun, openness, and being present are enough to transform your life. You will find yourself grateful, inspired, and enjoying living in a world you helped create when you can spread that to others.

For this to work, you do need to bring people along for the journey. This is a key distinction; it's a journey, not a destination. Just like all great travel adventures, there will be bumps, twists, and unexpected outcomes as you go. If you remain open and present, you will navigate them all with a smile.

We do encourage you to now go back and reread sections over. Find those areas you feel you need to ponder over more to achieve greater awareness through self-discovery. More importantly, stay the course when you start on your journey with these suggestions. You will know you are doing it right if it starts off being uncomfortable, then transitions to you being happier, followed by others mimicking how you deliver, empower, and inspire!

This book is part of The Ideation Emporium of Creativity™ series on Breakthrough Thinking™. The series is designed, much like The Ideation Emporium of Creativity™, to give you the exact things you need to move forward. There are two other books specific to this series. "The Five Letter F Word™" is a guide to setting prioritization, strategy, and communication throughout an organization to move faster than your competitors. "How Are Kids Innovating Faster Than You?™" gives a more detailed look at the innovation process that is designed so projects do not become stagnant and can reach the market as quickly as possible.

Thank you for your time, patience, and consideration.

For more innovation, downloads (the password is "innovate"), or to book a live Keynote or Workshop, contact us at www.IdeationEmporium.com.

ABOUT THE AUTHOR

As a former Executive Director of Marketing and having had responsibility for a portfolio of over $352 million, George understands the need for creativity in a business to drive the P&L. At one point, he successful launched 19 products in 14 months that indeed required different thinking that he is sharing with others. He even obtained a real estate license just to understand the needs of partners in that sector!

George's diverse background in business and science brings a lot of unique perspectives that others are missing. He now brings the extensive experiences of having worked in over 24 different industries to audiences worldwide using a unique approach to different sustainable Breakthrough Thinking™. Authentically sharing simple tools that anyone can use is the key to delivering real change fast.

Delivering a series of keynotes, seminars on creative ideation, and workshops on sustainable innovation, his unique tools help everyone grasp fun, exciting and tangible outcomes. He takes everyone's success seriously and embraces the needs of those he is helping.

George holds a black and multiple green belts in different academically accredited innovation processes such as Innovation, and Growth Systems, Green Belt for Growth and Value

Innovations. He also holds a Master of Science in Biology along with an MBA in Marketing and Management from Duquesne University. He also graduated with honors from the University of Pittsburgh with a Bachelor of Science in Biology.

George M. Nagle is the creator and journey leader for The Ideation Emporium of Creativity™. He is a single foster and adoptive parent to three amazing boys and actively mentors youth in the foster system. He is a master instructor in taekwondo and a volunteer coach with local youth sports.

He is the author of the Breakthrough Thinking™ series for business and life. The series includes "Miserable At Work? Why? You Don't Have To Be™," "The Five Letter F Word™," and "How Are Kids Innovating Faster Than You?™." George is also the author of a fictional spy series, "The Life We Lead: Ascending, Trials, and Sacrifice."

Made in the USA
Monee, IL
04 November 2026

67387906R00075